D1131252

Lecture Notes in Statistics

Edited by S. Fienberg, J. Gani, J. Kiefer, and K. Krickeberg

1

R. A. Fisher: An Appreciation

Edited by S. E. Fienberg and D. V. Hinkley

Springer-Verlag
New York Heidelberg Berlin

Editors
Stephen E. Fienberg and David V. Hinkley
Department of Applied Statistics
University of Minnesota
St. Paul, Minnesota 55108/USA

AMS Subject Classifications: 01A60, 01A70, 62-00, 62-03, 62CXX, 62D05, 62EXX, 62FXX

Library of Congress Cataloging in Publication Data

Main entry under title:

R. A. Fisher, an appreciation.

(Lecture notes in statistics)
1. Mathematical statistics—Addresses, essays, lectures. 2. Fisher, Ronald Aylmer, Sir, 1890-1962. I. Fienberg, Stephen E. II. Hinkley, D. V. III. Series.
QA276.16.R18 519.5 80-255

ISBN 0-387-**90476**-X Springer-Verlag New York Heidelberg Berlin
ISBN 3-540-**90476**-X Springer-Verlag Berlin Heidelberg New York

Printed in the United States of America.

9 8 7 6 5 4 3 2 1

Steward at First International Eugenics Congress, 1912

Fellow of the Royal Society, March 1929

Fisher at his desk calculator at Whittingehame Lodge, 1952

PREFACE

A very large number of the now-standard techniques of modern statistical theory and methods have their origins in the work of Sir Ronald A. Fisher. Yet our full professional debt to this great scientist often goes unacknowledged in modern courses on theoretical and applied statistics, and in modern textbooks. We believe that it is profitable for a student to consider the origins of modern-day statistics, in part because the original motivations for particular lines of development and philosophies have become clouded by time. This is particulary true of Fisher's work, which to an unfortunate extent has been superceded in the literature by work of a more mathematical nature. To study Fisher's statistical writings is to become enlightened, surprised -- and perhaps sometimes infuriated. To attempt to understand Fisher, and to argue over his writings, is to attempt a fundamental understanding the main threads of the fabric of statistical inference, and of modern statistical methods.

The present volume of lectures contains edited versions of presentations delivered as part of a faculty-student seminar and during a Special Lecture Series in the School of Statistics, University of Minnesota, during the Spring of 1978. The seminar was a cooperative effort, involving almost all of the faculty and graduate students in the School, and focussed on some of Fisher's major papers in the fields of estimation and statistical inference, design and analysis of experiments, contingency tables, distribution theory, discriminant analysis, and angular statistics. These papers are concentrated primarily in a 15 year period, 1920-1935, which saw giant strides in our subject and great debates involving Fisher and his contemporaries. The lectures were not intended as authoritative critiques of Fisher's work, but rather as introductions to and reviews of the key ideas in his writings. The same is true of the edited versions included here. The ordering of the lectures follows in a rough way the chronological ordering of the material being discussed.

In conjunction with the seminar the School of Statistics, with the support of the University of Minnesota Department of Concerts and Lectures and the College of Education, sponsored a Special Lecture Series under the general title: "R.A. Fisher: An Appreciation", now the title of this volume. Each lecture was given by a distinguished visitor, and three of these, by Joan Fisher Box, William Cochran, and David Wallace, fitted conveniently into the outline of material covered by the regular seminar series. Thus edited versions of these three lectures are included in the present volume. The remaining Special Lectures included a presentation on pre-Fisherian work by Stephen Stigler, and discussions of work that has built on Fisher's, by D.A.S. Fraser, I.J. Good, Marvin Kastenbaum, and Oscar Kempthorne.

Not only was Fisher one of the most productive and original statisticians of this century, but he also made fundamental contributions to genetics. There is an apocryphal story of a geneticist who, when introduced to a statistician several years ago, commented: "There's a well-known geneticist who I am told dabbles occasionally in statistics. Perhaps you have heard of him -- his name is Fisher". Actually, the extent of Fisher's writing in genetics (as well as related material in eugenics) is remarkable. One need only scan the complete list of Fisher's publications, reproduced here as an Appendix, to be impressed. While this material is clearly worthy of study, it lies beyond the boundaries of the seminar and Lecture Series, and is not discussed in the present volume.

Because so many of Fisher's publications are referred to in several lectures, we have reproduced his entire list of publications, as given in the Collected Papers of R.A. Fisher edited by J.H. Bennett (The University of Adelaide, South Australia: Coudrey Offset Press, 1974), at the end of this volume. References to Fisher's papers in the text are given in the style, CP 48, referring to paper No. 48 in the Collected Papers. All papers and books by others referenced in the lectures are listed in standard form at the end of each presentation.

The present volume of lectures has grown out of a detailed set of notes prepared by Christy Chuang, and typed by Karen Bosch. All of the lectures were subsequently rewritten, edited, and retyped with extensive assistance from Linda D. Anderson, who also prepared the final typescript. Without her skills and efforts the volume would not have been completed. The figures were drawn by Henri Drews and Sylvia Giesbrecht of the Department of Information and Agricultural Journalism at the University of Minnesota. Joan Fisher Box kindly supplied the photographs used. This work was supported in part by Office of Naval Research Contract N00014-78-C-0600 and National Science Foundation Grant MCS 7904558 to the Department of Applied Statistics, School of Statistics, University of Minnesota. Reproduction in whole or in part is permitted for any purposes of the United States Government.

We hope that the lectures here will help to introduce others to Fisher's work, so that yet another generation of statisticians can gain an appreciation of the relevant impact that his ideas still have on the way scientific research is conceived, carried out, and understood.

SEF
DVH
Minneapolis and St. Paul, Minnesota

November, 1979

CONTENTS

R.A. FISHER: SOME INTRODUCTORY REMARKS

David Hinkley

1. Introduction

R.A. Fisher was without question a genius. He brought statistical science from a collection of poorly understood, albeit useful, techniques toward a cohesive well-founded discipline. Along the way his strongly-held scientific views and somewhat difficult personality led to disputes, in some of which he was undoubtedly wrong. In almost no avenue of statistical science did he pursue ideas so exhaustively as to leave clear, unambiguous material. His strengths were intuition and inspiration, he spurned mathematical nit-picking. Compared to many of his contemporaries, his writing now appears lucid and surprisingly modern. However, much of Fisher's work is difficult, at times even exasperating, because of untied loose ends that are important to detailed understanding. In addition, many statistical results of practical importance were developed under Fisher's initial guidance, by junior colleagues such as Frank Yates.

Two outstanding bibliographic tours of Fisher's work are those by Yates and Mather (1963) and Savage (1976). Other useful biographical articles include those by Barnard (1963), Bartlett (1965), Kendall (1963), and Neyman (1967). During 1978 Joan Box, one of Fisher's daughters, published her extensive and highly informative biography, R.A. Fisher, The Life of a Scientist.

2. Early Background

Born in 1890 in North London, England, Fisher was encouraged from the early age of six to study astronomy and, later, mathematics. He attended Harrow, one of the foremost private schools in England, where he studied mathematics. During his school years he developed strong geometrical abilities, would solve problems in his head, and developed excellent mathematical intuition. In 1909, Fisher went to Gonville & Caius College, Cambridge, where he studied mathematics and physics -- including statistical mechanics and the theory of errors. There he read Karl

Pearson's "Mathematical Contributions to the Theory of Evolution" (1903), and he remained actively interested in evolution and genetics for the rest of his life.

Before we consider Fisher's contributions to statistics, it is useful for us to note what had been done by others up to 1909. An abbreviated list would include least squares estimation and the theory of errors (Gauss, Laplace and others), correlation (Galton), families of frequency curves and the fitting thereof, contingency tables and chi-square goodness-of-fit (Karl Pearson), components of variance (Edgeworth), asymptotic normal forms of likelihood (Laplace, Edgeworth), and Edgeworth's four papers on "probable errors". In 1908, Gossett, under the pseudonym "Student", published a derivation of the t-distribution. Substantive fields of application for statistics included economics, genetics, eugenics, physics, and astronomy. A large amount of Fisher's early theoretical work was addressed to the above list of topics, and much of this work will be discussed in the present set of lectures.

3. Outline of Fisher's Theoretical Work

While at Cambridge, Fisher published his first paper, "On an Absolute Criterion for Fitting Frequency Curves" [CP 1], in which least squares and the method of moments are criticized and the method of maximum likelihood is suggested. A simple distinction is drawn between moments estimation and maximum likelihood estimation, using the example of $\bar{X}$: since $(\bar{X} - \mu)^2$ has an average value of σ^2/n, the moment estimate of μ would be $\mu = \bar{X} \pm \sigma/\sqrt{n}$, whereas $\bar{X}$ is "the most probable value". Invariance of maximum likelihood under reparameterization is noted and contrasted with method of moments.

Sometime before the beginning of the Great War, Fisher was in contact with Karl Pearson and started work on the distribution of the (inter-class) correlation coefficient, r, which he reputedly derived in a week. The resulting paper [CP 4] represented a major breakthrough, and made splendid use of Fisher's geometric ability. Quite apart from the actual result, Fisher astutely realized the difference between r and ρ, which had been absent in earlier work on correlation. He also realized the futility of the probable error, unless the z-transform was employed. There is no doubt that by this date (1915) Fisher had already proved

"Student's" empirical result about t, but he did not see the need to publish the proof until 1925 [CP 43].

Apparently, Karl Pearson was slow to recognize Fisher's extraordinary talent, as he had been slow to recognize the importance of "Student's" work. Without consulting Fisher, Pearson set in motion a mammoth numerical investigation to obtain tables of the exact _density_ of r from Fisher's result. What particularly upset Fisher in the resulting 1917 _Biometrika_ publication (Soper, et al., 1917) was the imputation that he had invoked Bayes' theorem with a uniform prior for ρ; Pearson advocated an empirical prior which led to different numerical results. Fisher's calm, but definite, response appeared in _Metron_ in 1921 [CP 14], where he also established the distribution of the intra-class correlation, foreshadowing his derivation of the F distribution. Never again did Fisher publish in _Biometrika_.

During the years 1915-1919 Fisher was a school teacher. In 1919 he chose Rothamsted over University College, where Karl Pearson presided over the only group of academic statisticians. At Rothamsted Fisher had ample opportunity to pursue both statistical and genetic research. His first, and substantial, paper on inheritance-induced correlation had already appeared in 1918 [CP 9].

Before going to Rothamsted, Fisher had clear ideas about the current state of theoretical statistics, and was actively working on a unified treatment in which likelihood played an important part. A preview of his revolutionary 1922 paper [CP 18] appeared in an astronomy journal in 1920 [CP 12], where the estimation of observational accuracy (standard deviation) was discussed. Two methods for estimating dispersion were then in use by physicists, namely

$$n\hat{\sigma}_1 = \sqrt{\pi/2} \sum_{j=1}^{n} |X_j - \bar{X}|$$

and

$$n\hat{\sigma}_2 = \sum_{j=1}^{n} (X_j - \bar{X})^2 .$$

Fisher gave a geometrical derivation of their joint distribution for normal X, and noted the property of sufficiency for $\hat{\sigma}_2$. He concluded by commenting on the appropriateness of $\hat{\sigma}_1$ for long-tailed distributions, with a suggestion about testing for normality.

The 59-page 1922 paper [CP 18] introduced the fundamental concepts of consistency, efficiency, information, and likelihood. It criticized Pearson's method of moments on efficiency grounds. Later in 1922 [CP 19], Fisher handled the problem of chi-square and contingency tables. For the 2×2 table Pearson's chi-square had been used with three degrees of freedom by Pearson, but Fisher's geometric insight led to the correct single degree of freedom. This was not immediately accepted by others, especially by those associated with Pearson.

Further major theoretical developments were published in 1925 and 1934 [CP 42 and 108]; these two papers contained much of likelihood theory. The year 1934 saw a stormy presentation of Fisher's theoretical work at a meeting of the Royal Statistical Society [CP 124].

Fisher's curious views on probability crystallized in the 1930s with the advent of fiducial theory. A typical example of this theory is the inference from normal random samples that the population mean μ has the $N(\bar{X},\sigma^2/n)$ distribution. The theory floundered in a philosophical debate with Jeffreys, and received a jolt in connection with Behrens' problem of comparing two means.

I think that it will become clear as we read and discuss many of Fisher's primary statistical papers in subsequent presentations in this set of lectures that not only was he the founder of much of present-day statistical methodology, but that we can continue to learn new things by careful study of his writings.

References

Barnard, E.A. (1963). "Fisher's Contributions to Mathematical Statistics," Journal of the Royal Statistical Society, Series A, 126, 162-6.

Bartlett, M.S. (1965). "R.A. Fisher and the Last Fifty Years of Statistical Methodology," Journal of the American Statistical Association, 60, 395-409.

Box, Joan (1978). R.A. Fisher, The Life of a Scientist. New York: John Wiley & Sons.

Kendall, M.G. (1963). "Ronald Alymer Fisher, 1890-1962," Biometrika, 50, 1-16.

Neyman, J. (1967). "R.A. Fisher (1890-1962): An Appreciation," Science, 156, 1456-1460.

Pearson, K. (1903). "Mathematical Contributions to the Theory of Evolution. XII: On a Generalized Theory of Alternative Inheritance, with Special Reference to Mendel's Law," Philosophical Transactions of the Royal Society of London, Series A, 203, 53-87.

Savage, L.J. (1976). "On Rereading R.A. Fisher," The Annals of Mathematical Statistics, 4, 441-500.

Soper, H.E., A.W. Young, B.M. Cave, A. Lee, and K. Pearson (1917). "On the Distribution of the Correlation Coefficient in Small Samples. A Cooperative Study," Biometrika, 11, 328-413.

"Student" (1908). "The Probable Error of a Mean," Biometrika, 6, 1-25.

Yates, F. and K. Mather (1963). "Ronald Alymer Fisher," Biographical Memoirs of Fellows of the Royal Society of London, 9, 91-120.

This research was supported by National Science Foundation Grant MCS 7904558 to the University of Minnesota.

FISHER: THE EARLY YEARS

Joan Fisher Box

[Editors' note: The brief notes which follow were the outline for a Special Lecture by Joan Box. Since the lecture Mrs. Box's much-awaited biography of her father has been published under the title R.A. Fisher, The Life of a Scientist (Box, 1978).]

Fisher was 19 years of age when he went up to Cambridge and 29 when, in 1919, he took a job as statistician at the Rothamsted Experimental Station. There he was to find his feet in research, and soon become famous. This decade of his life gives an insight into the motivation and approach to science that were eventually to prove so rewarding, for what Fisher was doing and attempting then set the direction and laid the groundwork for all his later work. He distinguished himself in mathematical studies throughout school and college. As his college tutor wrote in 1919, "He could have been a first class mathematician had he stuck to the ropes, but he would not". As a mathematician he seemed to have disappointed early hopes; and he described himself as an "egregious failure" in two occupations after his college days.

His driving interest was in evolutionary biology and this directed his mathematical abilities onto new paths, paths not yet appreciated in 1919. Even at college, he had not stuck to the ropes. He saw the implications for the human race in Darwin's theory of evolution by natural selection, complemented by Mendel's genetical work, and he followed Francis Galton's concern that selection should tend constantly to improve the biological inheritance of man. His enthusiasm was infectious. In his second undergraduate year he was instrumental in the formation of the Cambridge University Eugenics Society. Speaking at the second monthly meeting of the undergraduate committee in November, 1911, he brought together Mendelian theory and biometrical method as the means of research in human heredity. His eugenic commitment presented him with both statistical and biological problems.

Research in both subjects began at once, though their relationship to each other was not obvious until the end of the period. In 1912 Fisher's first paper

[CP 1] appeared; there the method of maximum likelihood was introduced. This led to correspondence with W.S. Gosset ("Student"), which in turn resulted in his representation of a sample of n observations as a point in n-dimensional space, and consequently in his conception of the number of degrees of freedom and his derivation of "Student's" distribution. By geometrical argument he derived the general sampling distribution of the correlation coefficient in the summer of 1914. The paper [CP 4] was published in 1915 and contains a tentative suggestion of the z-transformation.

On the eugenic side, Fisher was the main speaker at the Second Annual Meeting of the Cambridge University Eugenics Society in 1912 where he proposed that the explanation for the rise and fall of civilizations lay in effects of social selection in a monied economy. He left college in 1913 and began work that was to continue for 20 years with the Eugenics Education Society of London. This brought him into close contact with the Society's president, Major Leonard Darwin, with whom a deep and enduring friendship resulted.

Fisher's eugenic commitment, like any genuine philosophy, was a way of life rather than a way of thinking. He volunteered for military service on the outbreak of war in August, 1914, and was very greatly disappointed by his rejection because of poor eyesight. He served his country by teaching in boys schools for five years, 1914-1919; it was a duty, not a pleasure to him. Disappointed in army service, Fisher developed ideas that farming was a eugenic way of life and an essential service to the nation. Its success depended on personal qualities rather than money, and it was the only occupation in which a large family would be an advantage. He married in 1917 and with his wife and sister-in-law, took a small holding and practiced subsistence farming. There, in the former gamekeeper's cottage, the first of Fisher's nine children was born in 1919. In the evenings extensive readings in the history of civilizations, with special attention to opinions or practices affecting the birthrate of different classes, were begun.

Despite his teaching job, farming enterprise, work for the Eugenics Society, and historical research, Fisher pursued the statistical research flowing from his paper on the distribution of the correlation coefficient. The Cooperative Study on the correlation coefficient, carried out under Karl Pearson's supervision at University College, was published in Biometrika in 1917 and made it clear that Fisher's work had been misunderstood and was disapproved of. Fisher completed a paper on the correlation to be expected between relatives on the supposition of Mendelian inheritance. This was communicated to the Royal Society in 1916, but later withdrawn because of the reports of the two referees. In this paper he introduced the word variance, and the techniques of the analysis of variance components as a means of distinguishing the contributions of ancestry, genotype, dominance, and other factors to the observed correlations. He thus showed human inheritance to be entirely consistent with Mendelian principles, a hypothesis that had been controversial since the beginning of the century. The paper was published elsewhere in October 1918 [CP 9], when Major Darwin insisted on sponsoring it. The job at Rothamsted also came to Fisher through a friend of Darwin's for whom he had done some statistical work during the war.

Fisher's work was difficult for his statistical contemporaries to assimilate, because of his inductive approach to the statistical interpretation of data and his central interest in biological variation. His statistical methods were at first equally difficult for the biologists to grasp, because of their mathematical sophistication. These same qualities made them great.

References

Box, J.F. (1978). R.A. Fisher, The Life of a Scientist. New York: John Wiley & Sons.

Soper, H.E., A.W. Young, B.M. Cave, A. Lee, and K. Pearson (1917). "On the Distribution of the Correlation Coefficient in Small Samples. A Cooperative Study," Biometrika, 11, 328-413.

DISTRIBUTION OF THE CORRELATION COEFFICIENT

Somesh Das Gupta

1. Introduction and Background

The extraordinary interest in the correlation coefficient is evidenced by the number of papers and pages in statistical publications dealing with the correlation coefficient (see Table 1).

Table 1: Number of Papers and Pages Dealing with the Correlation Coefficient Through 1966

Topic	Number of Papers	Number of Pages
Definition, Calculation, and Description	316	4651
Distribution of Correlation	71	889
Inference on Correlation	57	648

Source: Anderson, Das Gupta, and Styan (1972).

The following is a very brief history of the development of ideas about correlation prior to Fisher: Gauss derived the density function of the n-variate normal distribution from that of linear functions of independent normal variables in 1823 and 1826. August Bravais carried out work similar to Gauss in 1838 and 1846. Francis Galton introduced a numerical measure, r, which he termed "reversion" in a lecture at the Royal Institution on February 9, 1877 and later called "regression". The first use of the term "cor-relation" or "correlation" appeared in Galton's paper to the Royal Society on December 5, 1888, at which time correlation was defined in terms of deviations from the median. In 1892 Weldon referred to correlation as "Galton's function". Edgeworth replaced Galton's "index of co-relation" and Weldon's "Galton's function" by "coefficient of correlation" in 1892. K. Pearson and Sheppard gave the definition and calculation of correlation and its probable error in 1897. Subsequently, K. Pearson and Filon derived, in 1898, the standard deviation of r for large samples as $(1-\rho^2)/\sqrt{n}$, and took the distribution of r as normal unless ρ is very close to unity. "Student" derived the "probable error of a correlation coefficient" in 1908; Soper (1913) then used asymptotic expansions and normal approximations to better approximate the mean and standard deviation of r.[1]

The unsolved problem of finding the exact distribution of r for normal samples came to Fisher's attention via Soper's paper. His solution, reputedly obtained within one week, appeared in a remarkable paper [CP 4] which we shall examine in detail.

2. Fisher's Derivation of the Sampling Distribution of r [CP 4]

Let $(x_1,y_1),\ldots,(x_n,y_n)$ be a random sample from a bivariate normal distribution with means m_1,m_2, variances σ_1^2,σ_2^2, and correlation ρ. Define $\bar{x},\bar{y}$ as sample means, $ns_1^2 = \sum_1^n (x-\bar{x})^2$, $ns_2^2 = \sum_1^n (y-\bar{y})^2$, (Fisher uses μ_1^2 for s_1^2, μ_2^2 for s_2^2) and $nrs_1s_2 = \sum_1^n (x-\bar{x})(y-\bar{y})$. The problem is to find the distribution of r. The chance that n pairs fall within their specified elements is

$$\frac{1}{\left(2\pi\sigma_1\sigma_2\sqrt{1-\rho^2}\right)^n}\exp\left[-\frac{1}{2(1-\rho)^2}\sum_1^n\left\{\frac{(x-m_1)^2}{\sigma_1^2}-\frac{2\rho(x-m_1)(y-m_2)}{\sigma_1\sigma_2}+\frac{(y-m_2)^2}{\sigma_2^2}\right\}\right]dx_1dy_1\ \ldots\ dx_ndy_n. \qquad (1)$$

This can be interpreted as a sample density in 2n dimensions with $dx_1dy_1 \ldots dx_ndy_n$ as the probability element. The density part is already a function of $\bar{x}$, $\bar{y}$, s_1, s_2, r, but it is necessary to get the probability element also in terms of $d\bar{x}d\bar{y}ds_1ds_2dr$. Fisher did this by geometrical reasoning, which may be set out as follows.

Consider that subspace in the 2n-space of $(x_1,\ldots,x_n;\ y_1,\ldots,y_n)$ in which $\bar{x}$, $\bar{y}$, s_1, s_2, r are held fixed. Consider first the n-space of $x_1,\ldots,x_n$, illustrated in Figure 1 for n = 3.

---

[1]Detailed references for all pre-1900 articles on the correlation coefficient can be found in Anderson, Das Gupta, and Styan (1972).

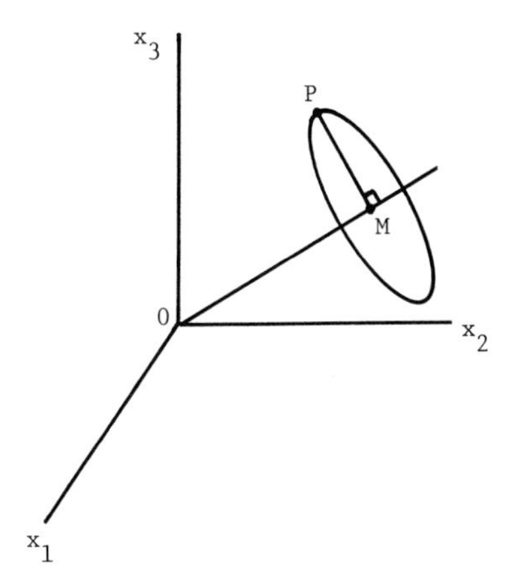

Figure 1. Representation of Sample Values of x for n = 3.

Let OM be the equiangular line $x_1 = x_2 = \dots = x_n$ with P the point $(x_1,\dots,x_n)$ and PM$\perp$OM. Then M is the point $(\bar{x},\dots,\bar{x}) = \bar{x}\,\underset{\sim}{e}$. If s_1 is also fixed, then we can only move the point P such that PM $= \sqrt{n}\, s_1$. Then P lies in an (n-1)-hypersphere with radius $\sqrt{n}\, s_1$; for n = 3, the hypersphere is the circle in Figure 1. If $\bar{x}$ is shifted by the amount $d\bar{x}$, and s_1 by the amount ds_1, then the volume element $dx_1 \dots dx_n$ is proportional to $s_1^{n-2} ds_1 d\bar{x}$ as in Figure 2. Similarly, for fixed $\bar{y}$ and s_2, Q = $(y_1,\dots,y_n)$ lies in the (n-1)-hypersphere with radius $\sqrt{n}\, s_2$ and center $(\bar{y},\dots,\bar{y})$.

The difficult part of the argument comes in relating the n-dimensional space of x and the n-dimensional space of y in the full 2n-dimensional space of the sample. Let $r = \cos\theta$ where θ is the angle between $(x_1 - \bar{x},\dots,x_n - \bar{x})$ and $(y_1 - \bar{y},\dots,y_n - \bar{y})$. If we define $x_i' = \bar{x} + k(y_i - \bar{y})$, for any k, then θ is also the angle between $(x_1 - \bar{x},\dots,x_n - \bar{x})$ and $(x_1' - \bar{x},\dots,x_n' - \bar{x})$. Let T = $(x_1',\dots,x_n')$. Now, fix the y-point Q and superimpose it in the x-space as T. Then the points on the (n-1)-hypersphere (on which P lies), which form an angle θ with MT, constitute an (n-2)-dimensional hypersphere. This is indicated in Figure 3 for the case n = 4.

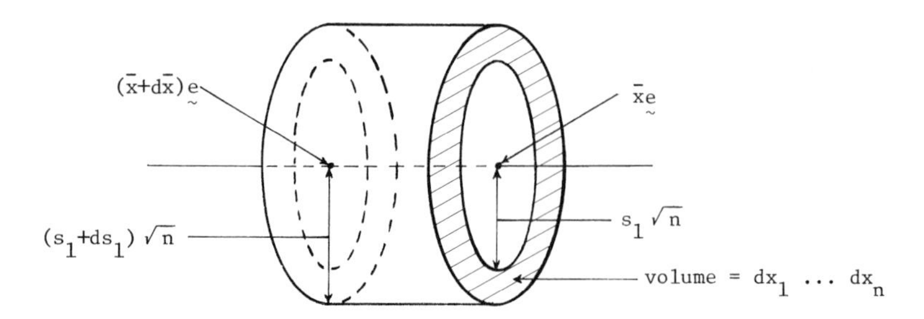

Figure 2. Volume element in n-space of x.

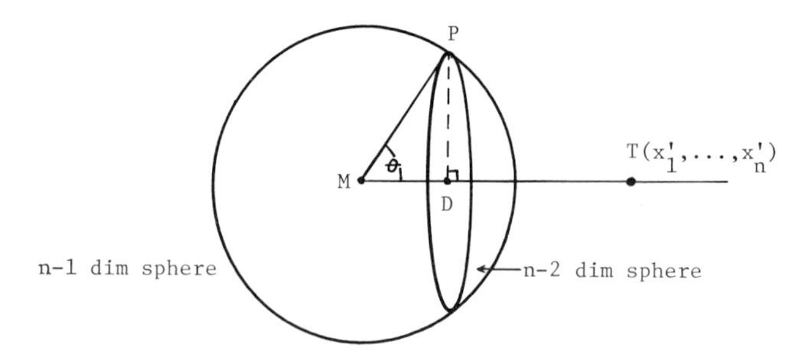

Figure 3. Relating the y-space and x-space.

Let D be the center of this (n-2)-dimensional hypersphere. Then PD $= \sqrt{n}\, s_1 \sin\theta = \sqrt{n}\, s_1 \sqrt{1-r^2}$. Figure 4 shows the effect of varying θ to θ + dθ. The volume due to changing θ to θ + dθ, $\bar{x}$ to $\bar{x} + d\bar{x}$ and s_1 to $s_1 + ds_1$ is proportional to $d\bar{x}(s_1\sqrt{1-r^2})^{n-3}(s_1 d\theta)ds_1$. So, the volume element in the full 2n-dimensional space is the volume product, i.e., proportional to $d\bar{y}s_2^{n-2}ds_2 d\bar{x}(s_1\sqrt{1-r^2})^{n-3}s_1 d\theta ds_1$. Since $\left|\frac{dr}{d\theta}\right| = \sin\theta = \sqrt{1-r^2}$, we have

$$\text{Volume element} \propto d\bar{x}d\bar{y}s_1^{n-2}ds_1 s_2^{n-2}ds_2(1-r^2)^{\frac{1}{2}(n-4)}dr. \qquad (2)$$

Now, combine expressions (1) and (2). Since $\bar{x}$ and $\bar{y}$ are independent of s_1, s_2 and r, we can integrate these out of the joint density function of $\bar{x},\bar{y},s_1,s_2,r$. Then the joint probability function of (s_1,s_2,r) is proportional to

$$\exp\left[-\frac{n}{2(1-\rho^2)}\left\{\frac{s_1^2}{\sigma_1^2} - \frac{2rs_1s_2}{\sigma_1\sigma_2} + \frac{s_2^2}{\sigma_2^2}\right\}\right] s_1^{n-2}s_2^{n-2}(1-r^2)^{\frac{1}{2}(n-4)}ds_1ds_2dr. \qquad (3)$$

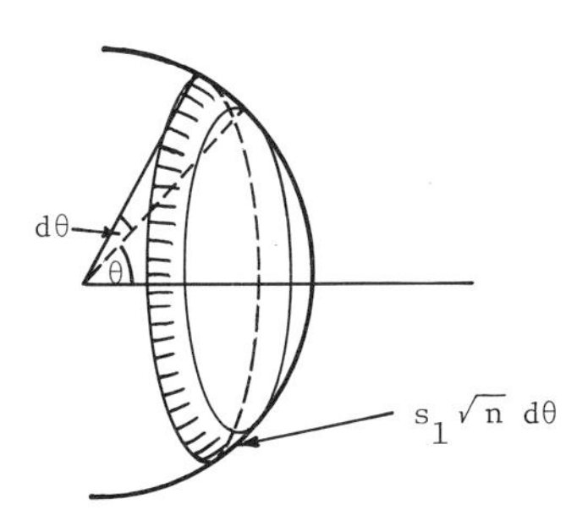

Figure 4. Effect in x-space of varying θ.

Let $\xi = \frac{s_1}{\sigma_1} \cdot \frac{s_2}{\sigma_2}$ and $e^z = \frac{s_1/\sigma_1}{s_2/\sigma_2}$. By a change of variables we get the joint probability density function of (ξ,z,r), and finally integrating (ξ,z) out we get the probability density element of r as

$$f(r) = C \int_{-\infty}^{\infty} dz \int_0^{\infty} \xi^{n-2} \, d\xi [e^{-n/(1-\rho^2)(\cosh z-\rho r)\xi}](1-r^2)^{\frac{1}{2}(n-4)} dr$$

$$= C \int_{-\infty}^{\infty} \frac{dz}{(\cosh z-\rho r)^{n-1}} (1-r^2)^{\frac{1}{2}(n-4)} dr. \qquad (4)$$

3. Further Reductions

Fisher's paper describes several representations other than expression (4) for the distribution of r. Perhaps the most important of these is based on a result due to Sheppard, which was mentioned to Fisher by Karl Pearson. Sheppard's result is that

$$\frac{1}{|\det \Gamma|^{\frac{1}{2}}} \int_0^{\infty}\int_0^{\infty} e^{-\frac{1}{2}u'\Gamma^{-1}u} \, du_1 du_2 = \cos^{-1}(-\gamma_{12}/\sqrt{\gamma_{11}\gamma_{22}}),$$

which implies that

$$\frac{n}{\sigma_1\sigma_2\sqrt{(1-\rho)^2}} \int_0^{\infty}\int_0^{\infty} e^{-\frac{1}{2}n(s_1^2/\sigma_1^2 - 2\rho r s_1 s_2/\sigma_1\sigma_2 + s_2^2/\sigma_2^2)/(1-\rho^2)} ds_1 ds_2$$

$$= \frac{\theta}{\sin\theta},$$

where $\cos\theta = -\rho r$.

Therefore $f(r) = c(1-r^2)^{\frac{1}{2}(n-4)} (\frac{\partial}{\partial r})^{n-2}\left(\frac{\theta}{\sin\theta}\right) dr$. Fisher used the special case $n = 2$ to get the constant c. The result is

$$f(r) = \frac{(1-\rho^2)^{n-\frac{1}{2}}}{\pi(n-3)!} (1-r^2)^{\frac{1}{2}(n-4)} \left(\frac{\partial}{\sin\theta\partial\theta}\right)^{n-2} \left(\frac{\theta}{\sin\theta}\right) dr.$$

Another representation for f(r) follows by noting that

$$\int_{-\infty}^{\infty} \frac{dz}{\cosh z + \cos\theta} = \frac{\theta}{\sin\theta}$$

implies

$$\int_0^\infty \frac{dz}{(\cosh z + \cos\theta)^{n-1}} = \frac{1}{(n-2)!}\left(\frac{\partial}{\sin\theta\partial\theta}\right)^{n-2} \frac{\theta}{\sin\theta} ,$$

so that writing $\cos\theta = -\rho r$ expression (4) becomes

$$f(r) = c(1-r^2)^{\frac{1}{2}(n-4)} \left(\frac{\partial}{\partial r}\right)^{n-2} \frac{\theta}{\sin\theta} \; dr.$$

4. The Relation Between r and "Student's" t Distribution

Let $t = \frac{r}{\sqrt{1-r^2}}$ and $\tau = \frac{\rho}{\sqrt{1-\rho^2}}$. Then the probability density function of t is

$$g(t) = c\left(\frac{\partial}{\sin\theta\partial\theta}\right)^{n-1} \frac{\theta^2}{2} \frac{1}{(1+t^2)^{\frac{1}{2}(n-1)}}$$

which when $\rho = 0$ is exactly the "Student's" t-distribution. In general,

$E(t) = \frac{n-2}{n-3}\tau$, and

$$\sigma_t^2 = \frac{1}{n-4}\{1 + \tau^2 + \frac{n-2}{(n-3)^2}\tau^2\}.$$

Fisher also calculates third and fourth moments of t and tabulates moments for different n and τ.

5. Maximum Likelihood Estimation of ρ

Fisher points out that since E(r) is less than ρ, one might estimate ρ by upward adjustment of r. But "This reasoning is altogether fallacious. ... It depends upon the choice of the particular variable [i.e., scale]". He goes on to say that the method of maximum likelihood (ML) does not depend on the choice of scale, and proceeds to derive the most probable value of ρ given r. This is the value of ρ which maximizes expression (4), and is therefore the solution $\tilde{\rho}$ to

$$\int_0^\infty \frac{dx}{(\cosh x - \rho r)^n}(r - \rho\cosh x) = 0.$$

Since $\cosh x > \tilde{\rho} r$, $r - \tilde{\rho}\cosh x$ must change sign. Therefore r must exceed the most probable value of ρ. The approximate solution is given by

$$r = \tilde{\rho}(1 + \frac{1-r^2}{2n}).$$

Note that $\tilde{\rho}$ is derived not from $(\bar{x}, \bar{y}, s_1, s_2, r)$, but from r alone. The usual MLE $\tilde{\rho}$ derived from the sampling distribution of $(\bar{x}, \bar{y}, s_1, s_2, r)$ is of course r. Fisher has used a "marginal" approach depending only on r and its sampling distribution, without explanation.

6. Transformation of r

Fisher notes that for ρ near $\pm$ 1, the distribution of r becomes extremely skew and changes form rapidly. This is one reason for working with t in the latter parts of the paper. In the concluding pages, he suggests the z-transform, $z = \tanh^{-1} r$, as possibly superior to t, in connection with reducing asymmetry of distribution and attaining approximate constancy of standard deviation for the working statistic. The suggestion, however, is not explored in any detail, and a full treatment appears later in 1921 [CP 14].

7. Pearson's Criticism

Following publication of Fisher's results, Karl Pearson set up a major cooperative study of the correlation, which seems to have proceeded with little attempt to actively involve Fisher. One particular aspect of the resulting publication (Soper, et al., 1917) that annoyed Fisher was the misunderstanding over maximum likelihood: Pearson and his co-workers criticized Fisher for using a Bayesian solution with the wrong prior distribution for ρ. Fisher's response to this criticism appears in his second major paper on correlation five years later [CP 14].

References

Anderson, T.W., S. Das Gupta, and G.P.H. Styan (1972). A Bibliography of Multivariate Statistical Analysis. Edinburgh: Oliver and Boyd.

Soper, H.E. (1913). "On the Probable Error of the Correlation Coefficient to a Second Approximation," Biometrika, 9, 91-115.

Soper, H.E., A.W. Young, B.M. Cave, A. Lee, and K. Pearson (1917). "On the Distribution of the Correlation Coefficient in Small Samples. A Cooperative Study," Biometrika, 11, 328-413.

"Student" (1908). "The Probable Error of a Mean," Biometrika, 6, 1-25.

This research was supported by U.S. Army Research Office Grant DAAG29-76-G-0038 to the University of Minnesota.

FISHER AND THE ANALYSIS OF VARIANCE

William G. Cochran

1. Introduction

The development of the analysis of variance and many of its applications is one of the main evidences of Fisher's genius. In this lecture I have described some of Fisher's papers on analysis of variance that particularly interested me. The first paper on this topic (with W.A. Mackenzie) appeared in 1923 [CP 32]. Two aspects of this paper are of historical interest. At that time Fisher did not fully understand the rules of the analysis of variance -- his analysis is wrong -- nor the role of randomization. Secondly, although the analysis of variance is closely tied to additive models, Fisher rejects the additive model in his first analysis of variance, proceeding to a multiplicative model as more reasonable.

Three years later Fisher had a fairly complete mastery of the main ramifications of the analysis of variance, including tests of significance, the role of randomization in providing estimates of error from the results of an experiment, the use of blocking, the analysis of split-plot experiments, factorial experimentation and confounding, as well as the null distribution of $z = \frac{1}{2} \ln F$ and its relation to the χ^2, t, and normal distributions.

Other works referred to in this lecture are a fairly intricate early example of confounding (the experiment was conducted in 1927), the analysis of covariance, work on the analysis of variance of Poisson or binomial data, and an attempt by Fisher to derive the distribution theory for a non-linear regression problem. These examples give an indication of Fisher's point of view and of the variety of problems that he tackled.

2. 1923: The First Paper on the Analysis of Variance [CP 32]

The first published paper with an analysis of variance was the analysis by Fisher and Mackenzie of the results of a $2 \times 12 \times 3$ factorial experiment on potatoes, which appeared in 1923. The factors were: (1) farmyard manure (dung) -- absent and present, (2) twelve standard varieties of potatoes, and (3) no potash, potassium sulphate, and potassium chloride. The experiment was laid out in a nested or split-split plot design.

The field was divided into two large pieces or blocks of land, one of which was given dung, the other not. Thus no estimate of the standard error for the average effect of dung was available, since this factor appeared in only single replication. Each block had 36 plots, containing the 12 varieties in three-fold replication, except that in the undunged block the variety Kitchener of Khartoum appeared only twice. This arrangement provides 47 degrees of freedom for testing the mean square for Varieties and the Varieties × Dung interactions. Fisher indicates particular interest in testing Variety × Manures interactions, since only in their absence could varietal comparisons made on a single manure, or manurial comparisons made on a single variety, be trusted.

No randomization was used in this layout. Following the procedure recommended at that time, the layout apparently attempts to minimize the errors of the differences between treatment means by using a chessboard arrangement that places different treatments near one another so far as is feasible. This arrangement utilizes the discovery from uniformity trials that plots near one another in a field tend to give closely similar yields. A consequence is, of course, that the analysis of variance estimate of the error variance per plot, which is derived from differences in yield between plots receiving the same treatment, will tend to overestimate, since plots treated alike are farther apart than plots receiving different treatments. Fisher does not comment on the absence of randomization or on the chessboard design. Apparently in 1923 he had not begun to think about the conditions necessary for an experiment to supply an unbiased estimate of error.

As regards the third factor, each plot contained three rows, one receiving potassium sulphate, one potassium chloride, and one no potassium. Thus the K comparisons and the VK, DK, and VDK interactions were made on a sub-plot basis. The analysis of variance that would now be considered appropriate for this experiment is shown in Table 1.

Table 1: Analysis of Variance for the Split-Split Plot Experiment

Source of Variation	d.f.
Blocks (D)	1
V	11
VD	11
Between-plot error	47
K	2
VK	22
DK	2
VDK	22
Within-plot error	94
Total	212

The Fisher-Mackenzie analysis of variance (Table 2) differs from Table 1 in two respects. They did not realize that the split-split feature necessitated two separate estimates of error. They combined the two estimates, giving 94 + 47 = 141 d.f. Also they did not distinguish D and K as separate factors, reporting 3 × 2 = 6 manures.

Table 2: Fisher and Mackenzie's Analysis of Variance

Source of Variation	d.f.	Mean Square	F
Manuring	5	1231.6	98.77
Variety	11	258.5	20.73
Deviations from summation formula	55	17.84	1.43
Variation between parallel plots	141	12.47	
Total	212	-	

This section of their paper is described as "analysis of variation on the sum basis" and in Table 2, the line "Deviations from summation formula", with 55 d.f., represents the Varieties × Manures interactions.

At that time Fisher had not published or tabulated the null distribution of F. For the Varieties × Manures interactions, his analysis gives F = 17.84/12.47 = 1.43 with 55 and 141 d.f. In testing this value, he uses an approximation to the null F-distribution that works well enough if the degrees of freedom exceed 20. He assumes that $\ln s_i$ is normal with mean $\ln \sigma_i$ and S.E. = $\sqrt{1/2v_i}$, where v_i is the number of degrees of freedom for s_i^2. It follows that on the null hypothesis

$\sigma_1 = \sigma_2$, $\frac{1}{2}$ ln F = 0.1788 has S.E. approximately $(1/110 + 1/282)^{\frac{1}{2}} = 0.1124$. The normal deviate 0.1788/0.1124 = 1.591, giving P = 0.056 in a one-tailed test. On this test the interactions mean square is not quite significant, although the authors note that the effects of both varieties and manuring are clearly significant.

Having concluded this section, the authors continue:

> The above test is only given as an illustration of the method; the summation formulae for combining the effects of variety and manurial treatment is evidently quite unsuitable for the purpose. No one would expect to obtain from a low yielding variety the same actual increase in yield which a high yielding variety would give; the falsity of such an assumption is emphasized by the fact that the expected values $(a + b - \bar{x})$ calculated on such an assumption, are often negative in the unmanured series. A far more natural assumption is that the yield should be the product of two factors, one depending on the variety and the other on the manure.

The remark about negative expected values may be an exaggeration. In this experiment, I found only one, under the additive model, that for the undunged rows receiving no potash, for which the expectation is (6.03 + 4.47 - 11.76) = -1.26. The logical point that a summation formula can lead to negative expected values is, however, sound.

3. The Product Formula -- Multiplicative Effects

If x_{ij} is the mean yield of the ith manure and the jth variety, the model for the product formula is

$$x_{ij} = a_i b_j + \varepsilon_{ij}.$$

The normal equations for estimating the a_i and b_j by the method of least squares are

$$\sum_j b_j x_{ij} = a_i \sum_j b_j^2, \qquad (1)$$

$$\sum_i a_i x_{ij} = b_j \sum_i a_i^2. \qquad (2)$$

Fisher notes that these equations may be solved by iteration. From first approximations to the a_i, equation (2) gives first approximations to the b_j, then equation (1) gives second approximations to the a_i, and so on until the solutions converge.

By this method he obtains a residual sum of squares of 847 on a single-observation basis.

As an alternative route, he notes that equations (1) and (2) give

$$\sum_i \sum_j a_i b_j x_{ij} = (\sum_i a_i^2)(\sum_j b_j^2) = \lambda. \tag{3}$$

From equation (3), it follows that the residual sum of squares is

$$\sum_i \sum_j (x_{ij} - a_i b_j)^2 = \sum_i \sum_j x_{ij}^2 - \lambda,$$

so that the value of λ is the reduction in sum of squares due to the product model and leads to the residual sum of squares.

Further, if equation (1), with k instead of j as the running subscript, is used to eliminate the a_i from equation (2), we get

$$\sum_i b_i (\sum_i x_{ij} x_{ik}) = b_j \sum_i a_i^2 \sum_k b_k^2 = b_j \lambda. \tag{4}$$

Hence if

$$c_{jk} = \sum_i x_{ij} x_{ik},$$

equation (4) shows that λ is a root of the determinantal equation

$$|c_{jk} - \lambda I| = 0.$$

By solving this equation Fisher finds the residual sum of squares to be 846.3, confirming the iterative solutions. The residual mean square is 846.3/55 = 15.39, slightly better than the residual mean square 17.8 from the additive model, as Fisher seems to have expected. Rather surprisingly, practically all of Fisher's later work on the analysis of variance uses the additive model. Later papers give no indication as to why the product model was dropped. Perhaps Fisher found, as I did, that the additive model is a good approximation unless main effects are large, as well as being simpler to handle than the product model.

4. The Distribution of $z = \frac{1}{2} \ln F$

In the next year, 1924, Fisher presented the null distribution of $z = \frac{1}{2} \ln F$ in an informative paper at the International Mathematical Congress at Toronto [CP 36]. This paper showed the relation between the z-distribution and the χ^2, normal, and t-distributions. I understand that the Toronto paper was not published until 1928.

Fisher does not give an analytical proof of the distribution, but remarks that he found it first when studying the distribution of the intraclass correlation coefficient. With n classes of size g, the covariance between members of the same class was estimated as

$$2 \sum_{i=1}^{n} \sum_{j=1}^{g} \sum_{k>g}^{g} (x_{ij}-\bar{x})(x_{ik}-\bar{x})/ng(g-1), \tag{6}$$

the number of products being $ng(g-1)/2$. With the variance of the x_{ij} estimated as

$$s^2 = \sum_{1}^{n} \sum_{1}^{g} (x_{ij}-\bar{x})^2/ng, \tag{7}$$

we get for the intraclass correlation coefficient

$$r = \frac{2 \sum_{i=1}^{n} \sum_{j=1}^{g} \sum_{k>j}^{g} (x_{ij}-\bar{x})(x_{ik}-\bar{x})}{(g-1) \sum_{i=1}^{n} \sum_{j=1}^{g} (x_{ij}-\bar{x})^2} . \tag{8}$$

Harris (1913) noted that with g large, the number of cross-product terms made r in equation (8) tedious to calculate. In finding a quicker method of calculating r, he established the link between r and the analysis of variance by proving that if SSB, SSW denote the sums of squares between and within groups,

$$\frac{SSB}{SSW} = \frac{(n-1)}{n(g-1)} F = \frac{\nu_1}{\nu_2} F = \frac{1 + (g-1)r}{(g-1)(1-r)} . \tag{9}$$

Equation (9) gives the relation between F and r. The quantities ν_1 and ν_2 are the degrees of freedom between and within classes, $\nu_1 = (n-1)$, $\nu_2 = n(g-1)$, so that $(g-1) = \nu_2/(\nu_1+1)$.

If c is a constant of integration, the density of r in the null case, which Fisher [CP 14] found in 1921, is

$$f(r) = c(1-r)^{(g-1)n/2 - 1} [1 + (g-1)r]^{(n-3)/2} \tag{10}$$

Substitution of $F = e^{2z}$ for r from expression (9), plus some tidying, gives the null distribution of z:

$$c' e^{\nu_1 z} dz(\nu_2 + \nu_1 e^{2z})^{-(\nu_1+\nu_2)/2} \tag{11}$$

This method of finding the distribution of z probably explains why the discussion of the analysis of variance in Statistical Methods for Research Workers [SMRW], which appeared in 1925, immediately follows intra-class correlation, coming in the same chapter.

5. 1925-26: Further Development of Analysis of Variance -- Randomization, Blocking, Confounding

When Statistical Methods for Research Workers [SMSW] appeared in 1925, it was evident that Fisher understood the function of randomization in experimentation. He writes in [SMSW, p. 248]: "The first requirement which governs all well-planned experiments is that the experiment should yield not only a comparison of different manures, treatments, varieties, etc. but also a means of testing the significance of such differences as are observed". And later "For our test of significance to be valid the differences in fertility between plots chosen as parallels must be truly representative of the differences between plots with different treatment; and we cannot assume that this is the case if our plots have been chosen in any way according to a prearranged system".

He notes that valid tests of significance are obtained if treatments are assigned to plots wholly at random (e.g., by shuffling or by a table of random numbers). Going further, he observes that the device of blocking, as in randomized blocks and the Latin square, combined with appropriate randomization, can give increased accuracy in the comparison of different treatments without sacrificing valid tests of significance.

It is also clear that by 1925 Fisher's earlier confusion about the nature of the experimental errors in split-plot experiments had been removed. In Statistical Methods for Research Workers [SMRW] he analyzes the dunged half of the 1923 2×12×3 split-split-plot experiment as a split-plot experiment with varieties in the plots and potash dressings in the sub-plots (rows), presenting the usual two errors -- one for comparison among plots and one for comparisons among sub-plots in the same plot.

A remarkable paper of 10½ small pages appeared in 1926 [CP 48], containing many of Fisher's ideas on the layout of field experiments. The paper, titled "The Arrangement of Field Experiments", opens by noting the recent interest of agriculturalists in the errors of field experiments, aided by study of the results of uniformity trials. It then raises the questions: What is meant by a valid estimate of error? When is a result statistically significant? The paper is, I think, the first in which Fisher puts forward his preference for a low standard of significance at the 5% level, noting that others may prefer a different level (2% or 1%). He then notes the vital role of replication in providing data from which to estimate errors, and repeats in more detail the argument for the importance of randomization in providing a valid estimate of error, including a description of randomized blocks and Latin square designs as arrangements that can give increased accuracy in the comparison of different treatments.

Then follows the advocacy of factorial experimentation, including the famous passage: "No aphorism is more frequently repeated in connection with field trials, than that we must ask Nature few questions, or, ideally, one question, at a time. The writer is convinced that this view is wholly mistaken. Nature, he suggests, will best respond to a logical and carefully thought out questionnaire; indeed, if we ask her a single question, she will often refuse to answer until some other topic has been discussed."

After giving an example of a factorial experiment, he ends this remarkable paper by pointing out that it will sometimes be advantageous to sacrifice information on some treatment comparisons by confounding them with certain elements of soil heterogeneity, in order to increase accuracy or other treatment comparisons considered more important.

6. An Early Example of Confounding

In 1927 Rothamsted conducted a 4×3×2 factorial experiment on barley, partially confounded in blocks of 12 plots, that presented problems both in the estimation of effects and in the analysis of variance. Fisher used this example twice: in an expository monograph [CP 90] with J. Wishart on analysis of the results of field experiments, and in a revised form in section 52 of his book Design of Experiments [DOE].

In a standard 4×3×2 factorial, with for example, 4 levels of K, 3 levels of N, and 2 levels of P, all 24 treatment combinations are distinct. In blocks of 12 plots we can include all 12 KN combinations and all 6 NP combinations, so that the KN and NP interactions need not be confounded with blocks. However, since there are 8 KP treatment combinations, the KP and KNP interactions must be partially confounded in blocks of 12 plots. In the best plan, one example of the two blocks in a replicate is as follows:

Block Ia

KNP	KNP	KNP	KNP
000	100	201	301
011	111	210	310
021	121	220	320

Block Ib

KNP	KNP	KNP	KNP
001	101	200	300
010	110	211	311
020	120	221	321

With any level of K, P must appear twice at the 1 level and once at the 0 level, or vice versa. In block Ia, P appears twice at the 1 level with K0 and K1, and twice at the 0 level with K2 and K3. The result is that in the KP interaction the comparison (K3 + K2 - K1 - K0)(P1 - P0) is partially confounded, while the two independent comparisons (K3 - K2 + K1 - K0)(P1 - P0) and (K3 - K2 - K1 + K0)(P1 - P0) are unconfounded with blocks Ia and Ib. The partially confounded comparison in KP can be estimated clear of blocks from the data in block Ia as

$$\tfrac{1}{4}\left[301 - \frac{(310+320)}{2} + 201 - \frac{(210+220)}{2} - \frac{(111+121)}{2} + 110 - \frac{(011+021)}{2} + 000\right].$$

The variance of this comparison from block Ia is therefore $6\sigma^2/16 = 3\sigma^2/8$, as against $2\sigma^2/6 = \sigma^2/3$ for the two components of KP that are unconfounded. In Fisher's terminology the units of information on the 3 comparisons in KP from block Ia are therefore $(3 + 3 + 8/3)/\sigma^2 = 26/3\sigma^2$ as against $9/\sigma^2 = 27/3\sigma^2$ if KP were unconfounded. Hence in this plan KP is confounded only to the extent 1/27. Similarly, the KNP interaction may be shown to have relative information 23/27, being confounded to the extent 4/27 or about 15%.

The extra complication in the 1927 Rothamsted barley experiment is that the factors were 4 types or qualities of N fertilizer -- sulphate of ammonia, chloride of ammonia, cyananide, and urea, 3 levels of N (0, 1, 2), and 2 levels of P (0, 1). Clearly, there is no difference among the four qualities of N at the lowest level of N where no N is given. Thus there are only 18 distinct treatment combinations in this example.

The partial confounding and the estimation of PQ and QNP must be reconsidered. In making quality comparisons we obviously want to omit the 0 level of N. The simplest method is to make them from the sum of the N1 and N2 levels, though, as we shall see, there is an argument for making them from the (N1 + 2N2) results, giving double weight to the N2 level.

The plan previously presented in blocks Ia and Ib is the one used in this experiment in two replicates, replacing K by Q. With the (N1+N2) levels, it looks at first sight as if the comparison (Q3 + Q2 - Q1 - Q0)(P1 - P0) is completely confounded. When we calculate this comparison from block Ia we get

$$S = -(310 + 320 + 210 + 220 + 111 + 121 + 011 + 021)/8,$$

all terms being - with none +. But we can get an unconfounded intra-block estimate of the relevant part of this QP comparison from

$$[S + 2(000 + 100 + 201 + 301)]/8.$$

In this way blocks Ia and Ib give an intra-block estimate of the comparison (Q3 + Q2 - Q1 - Q0)(P1 - P0) in QP. The variance of this estimate is $2(8 + 16)\sigma^2/64 = 3\sigma^2/4$. The two unconfounded components of QP give estimates from blocks Ia and Ib with variance $\sigma^2/4$. Fisher notes that the partially confounded component of QP has a variance 3 times that of the unconfounded components and is relatively poorly measured. He also notes that this component is the largest of the three numerically, which at first sight suggests that there was an unlucky choice of the component to confound. However, in the final analysis of variance, QP was not significant. The overall information on all three components from blocks Ia and Ib is $(4 + 4 + 4/3)/\sigma^2 = 28/3\sigma^2$, as against $36/3\sigma^2$ with no confounding. The relative information on QP is 7/9 -- not as good as 26/27 obtained when all 24 treatment combinations were distinct.

In experiments with qualitative and quantitative factors, another question of interest is: How should quality differences be measured? It is reasonable that the difference between two qualities at the N2 level should be twice that at the N1 level. In this event the comparison (N2-2N1) will show no QN interaction, making the main effects of Q larger when measured as (2N2 + N1) rather than as (N2 + N1). Fisher tried both methods, as shown in Table 3.

Table 3: Two Methods of Measuring Quality Differences in N

N Level	Totals (8 replicates) Sulphate	Chloride	Cyanamide	Urea
1	1524	1618	1615	1469
2	1693	2110	1607	1965

Analysis of Variance

	Using (N2 + N1) d.f.	Sum of Squares	Using (2N2 + N1) d.f.	Sum of Squares
Q	3	21,739	3	33,032
QN	3	23,332	3	12,039
Sum	6	45,071	6	45,071

Evidently (2N2 + N1) works better, giving larger Q main effects and smaller QN interactions.

The analysis of variance for the whole experiment is shown in Table 4. The linear response to N is large (74%), while the effects of P and the quality differences are both significant at the 1% level. None of the overall interactions is significant.

Table 4: Analysis of Variance of the 4×3×2 Experiment

Source of Variation	d.f.	Sum of Squares	Mean Squares	F
Blocks	3	12,216		
N linear	1	308,505	308,505	138.1**
N quadratic	1	7,420	7,420	4.9*
P	1	18,881	18,881	12.5**
Q	3	33,032	11,011	7.3**
QN	3	12,039	4,013	2.7
QP	3	7,870	2,623	1.7
QNP	3	6,634	2,211	1.5
Error	27	40,763	1,510	-

7. Analysis of Covariance

Fisher introduced the analysis of covariance in the fourth edition of Statistical Methods [SMRW] in 1932. The application was not to an actual experiment but to uniformity trial data Y of 16 plots of tea bushes, arranged in a 4×4 square. The covariate X consisted of the yields of the same 16 plots in the preceding growth period. If the experiment is laid out in a 4×4 Latin square, the error mean square is found to be 97.2. On the other hand, if the previous yields X are used to form four randomized blocks, maximizing the previous differences among blocks, the error mean square in the current period (Y data) is reduced to 40.7. Use of the Latin square plan plus the covariance adjustments on previous yields does still better, reducing the error mean square to 15.1, a very marked gain in accuracy over the original Latin square.

Since there are no treatments in this example, a description of the F-test of the adjusted treatment means was postponed until the 5th edition of Statistical Methods [SMRW, 1934].

8. Analysis of Variance for Poisson and Binomial Data

In 1940 Cochran published a paper on the analysis of variance for data that follow the Poisson or binomial laws. With Poisson data, the method recommended at that time by Bartlett (1936) was to transform the data to square roots, and perform the analysis of variance in the square root scale. The idea was that in the square root scale the error variance should be approximately constant, so that the pooled error in the analysis of variance could be used, and there was a little evidence that the square roots were not far from normally distributed. With binomial data, Bliss (1937) recommended an analysis in the angular arcsine $\sqrt{p}$ scale, for similar reasons. Later, various adjustments to these transforms were introduced that were claimed to give a more constant variance on the transformed scale.

With Poisson data, Cochran's interest was as follows. Suppose one could assume that effects were additive in the square root scale, but wanted maximum likelihood estimates of the treatment effects, how would these differ from the estimates obtained in the recommended analysis of variance in the square root scale? With randomized blocks data, for example, let $\mu_{ij} = E(x_{ij})$. With an additive model in the square root scale, we have

$$\eta_{ij} = \mu_{ij}^{\frac{1}{2}} = \mu + \alpha_i + \beta_j. \tag{12}$$

Also, with Poisson data the log likelihood is

$$L = \sum_i \sum_j x_{ij} \ln \mu_{ij} - \mu_{ij}; \quad \frac{\partial L}{\partial \mu_{ij}} = \sum_i \sum_j \frac{(x_{ij} - \mu_{ij})}{\mu_{ij}}. \tag{13}$$

Hence, the normal equation $\partial L / \partial \alpha_i = 0$ for the parameter α_i is

$$\sum_j \frac{2}{\sqrt{\mu_{ij}}} (x_{ij} - \mu_{ij}) = 0. \tag{14}$$

Let $y_{ij} = \sqrt{x_{ij}}$. By a first-term Taylor expansion,

$$x_{ij} - \mu_{ij} \cong 2\sqrt{\mu_{ij}} \, (y_{ij} - \eta_{ij}). \tag{15}$$

Substituting from (15) back into expression (14) gives for the normal equation for α_i in the square root scale (with ML estimation),

$$4 \sum_j (y_{ij} - \eta_{ij}) = 0. \tag{16}$$

But expression (16) is the standard equation of estimation for α_i in an analysis of variance in the square root (y_{ij}) scale. Equation (16) is only approximate, depending as it does on a first-term Taylor expansion. Fisher's device, which Cochran used, is to turn this approach into a series of successive approximations to the ML estimates. Define an adjusted transform, y'_{ij}, so that equation (15) becomes correct. That is, let

$$y'_{ij} = \hat{\eta}_{ij} + \frac{(x_{ij}-\hat{\mu}_{ij})}{2\sqrt{\hat{\mu}_{ij}}} = \hat{\eta}_{ij} + \frac{(x_{ij}-\hat{\eta}_{ij}^{\,2})}{2\hat{\eta}_{ij}} . \qquad (17)$$

Equation (17) is used iteratively. From an analysis of variance in square roots, estimate the $\hat{\eta}_{ij}$. From (17), get a second approximation to the y'_{ij}. Analyze these y'_{ij} and get third approximations to the y'_{ij} from (17), continuing until the $\hat{\eta}_{ij}$ converge to ML estimates.

From a number of examples with Poisson or binomial data in which I found ML estimates of the treatment effects in this way, my summary impressions were: (i) the ML estimates usually differed little from those given by an analysis of variance in the square root or angular scale, and (ii) the adjustments to the transforms needed to give ML estimates were rather similar to those recommended to give more nearly constant variance in the transformed scale.

To illustrate what I mean, Table 5 gives results from the analysis of a 5×5 Latin square with Poisson data. The table shows (i) the observations, (ii) the square roots of the observations, (iii) the adjusted transforms that give the ML estimates by Fisher's method. Where the same observation, e.g., 4 occurred several times in the experiment, the mean of the adjusted transforms is shown in Table 5, (iv) the values $(\sqrt{x} + \sqrt{x+\frac{1}{2}})/2$. This adjusted transform is similar to the adjusted transform $(\sqrt{x} + \sqrt{x+1})$ recommended for better variance stabilizing by Freeman and Tukey (see Mosteller and Youtz (1961)), except that the adjustment in (iv) is somewhat milder. Note how closely the ML adjusted transforms in (iii) agree with the values $(\sqrt{x} + \sqrt{x+\frac{1}{2}})/2$ suggested in (iv) for improved variance stabilization.

In working these examples I also noticed on several occasions that data which appeared at first sight to be Poisson or binomial had extraneous variation present, as revealed by a χ^2 test. With such data I formed the judgment that analysis of the transforms or adjusted transforms might be justified as a working approximation, but Fisher's method could not be regarded as a more exact analysis, since the data were not Poisson or binomial. I formed the opinion that adjusted transforms might serve as a substitute for Fisher's ML solutions.

Table 5: Two Sets of Adjusted Square Roots with Poisson Data

(i)	x	0	1	2	3	4
(ii)	$\sqrt{x}$	0	1	1.41	1.73	2.00
(iii)	ML adj.	0.54	1.05	1.48	1.77	2.05
(iv)	$(\sqrt{x}+\sqrt{x+\frac{1}{2}})/2$	0.35	1.12	1.50	1.80	2.06
(i)	x	5	6	8	9	17
(ii)	$\sqrt{x}$	2.24	2.43	2.82	3.00	4.12
(iii)	ML adj.	2.25	2.47	2.86	3.04	4.17
(iv)	$(\sqrt{x}+\sqrt{x+\frac{1}{2}})/2$	2.29	2.50	2.87	3.04	4.15

In 1954, Fisher [CP 254] presented a paper in which he repeated his method of finding maximum likelihood estimates in the analysis of non-normal data by transforming the data. He then remarked that some authors seemed to think erroneously that the objective of the transformation was to find a constant error variance in the transformed scale. He noted that with Poisson data, Bartlett had proposed the adjusted scale $\sqrt{x+\frac{1}{2}}$ while Anscombe had proposed $\sqrt{x+3/8}$, neither author realizing that in Fisher's view these adjustments were a waste of time. He brought Cochran into the act by noting that while Cochran presented the ML approach, he did not "totally disavow" the analysis of transformed data adjusted for variance stabilization.

In retrospect I agree that Fisher's approach is superior. It specifies a definite mathematical model, and uses maximum likelihood estimation, recognized as preferable to least squares estimation with non-normal data. My results were that analysis of variance on the transformed scale, with or without variance-stabilizing adjustments, agreed closely with the ML estimates, and was a good working method, particularly when extraneous variation is present so that the assumptions leading to Fisher's ML solutions do not apply.

9. An Example of Non-Linear Regression

With regressions non-linear in some of the parameters, examples in which something approaching the null distribution of F can be found are rare. An example given by Fisher [CP 163] is the harmonic regression

$$y_x = \mu + \alpha \cos(\theta x) + \beta \sin(\theta x) + \varepsilon \qquad (18)$$

for $x = 1,2,\ldots,n = 2s + 1$. This regression is linear in μ, α, β, but non-linear in θ. With a linear regression the parameter μ is estimated by the mean $\bar{y}$, and in the null case the three parameters α, β, and θ would be expected to account for about a fraction of 3/2s of the sum of squares $\Sigma(y-\bar{y})^2$.

Fisher notes that the $n = 2s + 1$ observations may be expressed as $(2s+1)$ orthogonal linear functions, namely the mean $\bar{y}$ and the harmonic functions $\Sigma a_x y_x$, $\Sigma b_x y_x$, where

$$a_x = \sqrt{\frac{2}{2s+1}} \cos\left(\frac{2\pi p x}{2s+1}\right) \qquad b_x = \sqrt{\frac{2}{2s+1}} \sin\left(\frac{2\pi p x}{2s+1}\right), \qquad (19)$$

$p = 1,2,\ldots,s$. For example, with $n = 5$, $s = 2$ the four cos and sin terms are shown in Table 6 (without the multiplier $\sqrt{2/(2s+1)}$).

Table 6: Harmonic Terms ($n = 5$, $s = 2$)

p \ x		1	2	3	4	5
1	cos	.309	-.809	-.809	.309	1
1	sin	.951	.588	-.588	-.951	0
2	cos	-.809	.309	.309	-.809	1
2	sin	.588	-.951	.951	-.588	0

These terms are mutually orthogonal and are orthogonal to the mean $\bar{y}$. Therefore if $\alpha = \beta = 0$, so that the data have no harmonic trend, the quantities

$$ss_p = [\sum_x (a_x y_x)]^2 + [\sum_x (b_x y_x)]^2 \qquad (20)$$

are independently distributed as $\chi^2\sigma^2$ with 2 d.f. for any fixed value of $p = 1,2,\ldots,s$. The sum of squares (20) is the reduction in $\Sigma(y-\bar{y})^2$ due to fitting the model (18) with any given p and accompanying a_x and b_x. Of the s values of p, suppose we select the value that gives the greatest reduction in $\Sigma(y-\bar{y})^2$. Fisher

[CP 75] had shown previously that for this value of p, the ratio $g = ss_p/\Sigma(y-\bar{y})^2$ is distributed so that the probability of exceeding any value g is

$$P = s(1-g)^{s-1} - \frac{s(s-1)}{2}(1-2g)^{s-1} + \ldots + (-)^{k-1}\frac{s!}{(s-k)!k!}(1-kg)^{s-1},$$

the series stopping at the largest integer k less than 1/g. The upper 5% and 1% levels of the distribution of g have been tabulated by Eisenhart, Hastay, and Wallis (1947).

Fisher notes that if we fit the optimum θ in the model (18), it will give a reduction in the sum of squares at least as great as the best of the p's. Hence the table of g can be used to test the null hypothesis $\theta = 0$ under the model (18). The verdict of the test can be trusted if it finds a significant harmonic component, but if it fails to reject the null hypothesis, there might still be a value of θ, better than the best of the p's, that would give a significant result.

Fisher's work also shows that this is an example in which linear theory cannot be trusted. As s increases, the ratio $g = \max ss_p/\Sigma(y-\bar{y})^2$ does not tend when s is large to 3/2s, but to the larger value $(\gamma + \ln s)/s$ when s is large, where γ is Euler's constant. Similarly, the use of the F-ratio for 3 and (2s-3) d.f. for a joint test of $\hat{\alpha}$, $\hat{\beta}$, and $\hat{\theta}$, as linear theory would suggest, can be badly wrong. Some 5% F values for 3 and (2s-3) d.f. are shown below, along with the 5% F values from the g table, which are themselves a little low for testing the optimum $\hat{\theta}$.

Table 7: Comparison of F values

n	11	21	41	61
$F_{3,2s-3}$	4.35	3.20	2.86	2.52
F from g-distribution	5.04	4.54	4.57	4.69

The linear approximation becomes badly wrong as n increases.

References

Bartlett, M.S. (1936). "The Square Root Transformation in the Analysis of Variance," Journal of the Royal Statistical Society, 3, 68-78.

Bliss, C.I. (1937). "The Analysis of Field Experimental Data Expressed in Percentages," Plant Protection (Leningrad), 67-77.

Cochran, W.G. (1940). "The Analysis of Variance when Experimental Errors Follow the Poisson or Binomial Laws," The Annals of Mathematical Statistics, 11, 335-347.

Eisenhart, C., Hastay, M.W. and W.S. Wallis (1947). Techniques of Statistical Analysis. New York: McGraw-Hill.

Harris, J.A. (1913). "On the Calculation of Intra-Class and Inter-Class Coefficients of Correlation from Class Moments when the Number of Possible Combinations is Large," Biometrika, 9, 446-472.

Mosteller, F. and C. Youtz (1961). "Tables of the Freeman-Tukey Transformations for the Binomial and Poisson Distributions," Biometrika, 48, 433-440.

RANDOMIZATION AND DESIGN: I

Norton Holschuh

1. Introduction

This first lecture on R.A. Fisher's work in the area of design and analysis of experiments focusses primarily on the principles explained in his landmark 1935 book The Design of Experiments [DOE]. To highlight important issues in randomization theory, Fisher's initial controversy with Neyman and his rarely-heard opinion on the problem of undesirable random assignments will also be briefly reviewed.

2. The Design of Experiments [DOE]

The Design of Experiments [DOE] can be usefully divided into two parts. Fisher expresses his view on making inductive inferences from experimentation, and on the role of the structure of an experiment in that process in the first three chapters. He argues the necessity of randomization for performing valid tests of significance and obtaining valid estimates of experimental error. He also describes the principal ways of increasing the sensitivity of an experiment. All of these ideas are illustrated by two simple examples: (1) the somewhat artificial "lady tasting tea" experiment, and (2) an analysis of data from a Charles Darwin experiment. The remainder of the book, in general, applies these principles of significance testing and estimation to specific situations, including factorial experiments and such variance reduction schemes as blocking, confounding, and the use of concomitant measurements. It is the first three foundational chapters that I discuss here.

In his introduction, Fisher states the essential fact that the analysis or interpretation of an experiment is inextricably connected to the structure or design of the experiment. Consequently, if one is to expand scientific knowledge through experimentation, certain valid principles must exist. In particular the uncertainty inherent in inductive inference can be made rigorous by probabilistic reasoning. However, Fisher rejects the notion of using Bayes' theorem to make probabilistic inferences about the hypotheses given the results of an experiment. His major complaint is that such reasoning evidently forces one to "... regard mathematical probability, not as an objective quantity measured by observable frequencies, but as measuring merely psychological tendencies, theorems respecting which are useless for scientific purposes" [DOE, p. 6]. Furthermore, Fisher claims that properly designed experiments, as described in the remainder of the book, yield valid inferences without reference to Bayes' theorem.

In the second chapter, Fisher explains the notions of adequate design, tests of significance, randomization, and sensitivity, all through the lady tasting tea example. The lady claims that when presented with a cup of tea with milk she can determine whether tea was added first (TM) or milk first (MT), at least to some degree better than guessing. Fisher's experiment to test this claim involves presenting the lady with four portions of TM and four portions of MT placed in random order into eight cups. The lady is told the complete design of the experiment and must choose four cups as TM and four cups as MT. Fisher's first point is that the interpretation of the results and the logical basis for the interpretation must be decided beforehand. If one assumes that the lady cannot discriminate, then it might be reasonable to believe that she would, with equal probability, choose any combination of four to label TM. That this belief, which is most likely false, does indeed lead to the correct distributional assumptions is argued later. If this is the case, however, the expected frequency of number of correct selections is given in Table 1.

Table 1. Assumed Frequencies If No Discriminating Ability

Number Correct	Frequency
0	1/70
2	16/70
4	36/70
6	16/70
8	1/70

The best possible result for the lady is labelling all 8 cups correctly. Since this occurs with the small frequency of 1/70 under the assumption of no discriminating ability, the result could be considered significant evidence that the assumption of no ability is not true. The next best result of labelling 6 cups correctly would occur with frequency 16/70, so under the assumption of no ability the probability of labelling 6 or more cups correctly is 17/70. Since 17/70 is a relatively large frequency, this result is not considered to be nearly as much evidence against the assumption of no ability as is labelling all 8 cups correctly. The other possible results are interpreted similarly. This method of interpretation is termed a test of significance.

As explained by Fisher, a test of significance has the following features. There is some condition, referred to as the "null hypothesis", against which the experiment is designed to obtain evidence. Some function of the results admits a reasonable ordering of all possible results in terms of evidence against the truth of the null hypothesis. This null hypothesis is sufficiently exact to give this function a well-defined probability distribution when the null hypothesis is true. The level of significance of a result is the probability under the null hypothesis of observing a result with at least as much weight against the null hypothesis. Finally and most importantly, there is some logical argument for the correctness of this distribution.

Fisher stresses two points in regard to significance tests: (1) Through experimentation a null hypothesis can be disproved but never proved. Evidence can overwhelmingly accumulate against the null hypothesis, but lack of such evidence does not imply that the null hypothesis is true. Other hypotheses could equally well have produced the results. (2) Similarly, no one experiment can disprove the null hypothesis, since under the null hypothesis every result considered occurs with some positive probability. Consequently, it is the procedure and not any single result that determines experimental proof. Fisher states: "In relation to the test of significance, we may say that a phenomenon is experimentally demonstrable when we know how to conduct an experiment which will rarely fail to give us a statistically significant result" [DOE, p. 14].

As Fisher notes, the distribution under the null hypothesis must have some objective basis. The surprising fact is that presentation of the cups in random order logically insures that all possible labellings of the portions are equally likely. The random order is generated by some physical procedure, such as coin flipping, with probabilistic properties that are adequately accepted from a frequency viewpoint. Under the null hypothesis the lady has no discriminating ability, so that her choices are based on factors that do not depend on whether any particular cup contains a TM or an MT. Thus one can assume that all environmental conditions with the exception of the order of the portions are fixed. If the lady has no discriminating ability the frequencies of her labellings under those environmental conditions are also conceptually fixed. Since the physical randomization is independent of the environmental conditions, and the frequency distribution generated by the procedure is accepted as uniform over all possible orderings, there is an objective basis to claim the distribution of correct labellings is as listed in Table 1. This is true regardless of the lady's conceptual frequencies for a given set of environmental conditions. Perhaps one way to induce this distribution without randomization would be to make the environmental conditions for each portion identical. Fisher points out that this is impossible in this case and virtually all others.

If the lady does have discriminatory ability, however, there is no guarantee that the experiment will, with any substantial probability, yield significant results. The sensitivity of the experiment to deviations from the null hypothesis must come from other aspects of the design. Fisher illustrates the three general ways for improving the sensitivity of an experiment. First, one can increase the size of an experiment. In this instance such an increase allows the lady to achieve a high level of significance and yet have one or more mistaken labellings. Second, one can refine the experimental technique. The environmental conditions for each portion can be made as uniform as practical. This reduces environmental obstructions to the lady's discriminatory ability. For example, a clearly avoidable obstruction would be having only the first four cups contain sugar. But Fisher also emphasizes that one should not go to excessive lengths to make the conditions uniform. The experiment is valid regardless, and one might increase sensitivity more economically by increasing the size of the experiment. Finally, one can change the structure of the experiment. Fisher's example is flipping a fair coin to assign treatments (TM or MT) to each cup. With this procedure there is probability 1/256 of labelling all correctly and probability 8/256 of labelling 7 cups correctly and 1 cup incorrectly. Hence she can make one mistake and still obtain fairly high significance.

In his third chapter, Fisher proceeds to apply the principles illustrated in the tea-tasting experiment to an analysis of a biological experiment. The data considered arose from part of an experiment by Charles Darwin to determine the effects of self- and cross-fertilization on the heights of different varieties of plants. For a specific variety, self- and cross-fertilized plants were grown in pairs. Environmental conditions were equalized as nearly as possible. The members of a pair were always planted in the same pot. The results for the variety *Zea mays* are presented in Table 2.

Table 2. Heights for 15 *Zea mays* Pairs (1/8 inch).

	Pair No.	Cross-Fertilization	Self-Fertilization	Difference C-S
Pot 1	1	188	139	49
	2	96	163	-67
	3	168	160	8
Pot 2	4	176	160	16
	5	153	147	6
	6	172	149	23
Pot 3	7	177	149	28
	8	163	122	41
	9	146	132	14
	10	173	144	29
	11	186	130	56
Pot 4	12	168	144	24
	13	177	102	75
	14	184	124	60
	15	96	144	-48

The important idea illustrated by this example is the notion of blocking or, in this instance, matching. Fisher observes that Darwin's common sense consideration of making conditions as uniform as possible for each pair was certainly correct. In fact blocking is a solution to the problem of obtaining adequate replication without being harmed by the usually accompanying increased variability of the experimental material. The contrast of interest, the difference in height between cross- and self-fertilized plants, is estimated within each pair so that any amount of variability among the group of pairs is of no consequence. The precision of the estimated contrast is affected only by the uniformity of environmental conditions within each pair and, of course, the amount of replication of pairs.

By the above principles the structure of this experiment should dictate the analysis. Specifically, the estimate of the error inherent in the estimate of the contrast should be derived from only the sources of variability affecting that error. To remove the variability among pairs, which does not affect the error of the estimated contrast, Fisher analyzes the 15 differences shown in the last column of Table 2. He argues that biological data often follow the normal "theory of errors" and hence applies the single sample t-test to the fifteen differences, conveniently ignoring the fact that sets of pairs in the same pot could likely be correlated. In any case he obtains a t-value of 2.148. This yields a (2-sided) significance level of 4.97%. For an example of an incorrect analysis, caused by ignoring the structure of the experiment, Fisher produces Francis Galton's rearrangement of the data given in Table 3.

Table 3. Galton's Rearrangement

Cross-fertilization	Self-fertilization	Difference C-S
188	163	25
186	160	26
184	160	24
177	149	28
177	149	28
176	147	29
173	144	29
172	144	28
168	144	24
168	139	29
163	132	31
153	130	23
146	124	22
96	122	-26
96	102	-6

Galton rearranged the data evidently to show the "statistical regularity" of the two sets of measurement and thus the "reliability" of their means. Fisher performs a t-test on the third column of differences obtaining a t-value of 5.171, an extremely large result under the null hypothesis. This arises from the low estimate of error, which in turn is a result of pairing the ordered values of the cross-fertilized with the ordered values of the self-fertilized. This artificial pairing eliminates elements of error that exist in the structure of the experiment from the estimate of error affecting the differences.

While the structure of the experiment determines the sources of error, there remains the question of the validity of the estimate of the error. Fisher argues that by making an independent, random assignment of relative positions for each pair one insures that the usual estimate of error for the t-test is valid. Environmental conditions that favor a particular position in a pair will, with equal frequency over all possible assignments, benefit each method of fertilization. Since the random assignments are independent for each pair, straightforward calculations show that the usual estimate of error is indeed valid in terms of the conceptual distribution of all possible assignments. So, as in the tea-tasting example, the probability distribution in terms of which the results are to be viewed, is that generated by the nature of the random assignments. According to Fisher, the lack of a random assignment of positions in each pair was Darwin's only mistake in an otherwise well designed experiment.

Had Darwin randomized his assignments, one could perform a significance test, the validity of which depends only on the randomization distribution. Fisher refers to this as a "test of a wider hypothesis" since it does not depend on the normality assumptions of the t-test. Under the null hypothesis of no difference in height between the populations of cross- and self-fertilized plants, the differences observed are due to environmental differences in the positions, measurement error, and error in sampling from the supposedly identical populations of cross- and self-fertilized plants. Ignoring the problems of measurement error and sampling error, which Fisher does, each observed difference occurs with a positive or negative sign as a function only of the assignment of positions in the pair. Thus one knows the conceptual results of all possible 2^{15} assignments; and, because of the randomization procedure, the assumption that all could have occurred with equal probability is valid. Consequently, the frequency distribution of any function of the data is known subject to its calculation for all possible assignments. The statistic which Fisher chooses to order the possible results is the absolute value s of the sum of the differences, a monotonic function of the absolute t-statistic. The larger is the value of the statistic, the more evidential weight there is against the null hypothesis. For Darwin's particular assignment the value is s = 314 eighth inches. The calculations carried out by Fisher show that 1670 possible assignments yield a larger value and 56 yield the same value of s. The significance level is thus $(1670 + 56)/(2^{15})$ or approximately 5.267 percent. Fisher notes how close this is to the level given by the t-test, although the comparison is meaningless in a strict sense, since Darwin did not perform the assumed randomization procedure. However, it is now known that for the common experimental designs, the null distributions under all possible randomizations of the normal theory test statistics (F-tests and t-tests) are reasonably well approximated by their distributions under the usual normality assumptions. Since finding the randomization distribution is usually time consuming, the distribution derived from normality assumptions is often used as an approximation.

In a final section to Chapter 3, added in later editions, Fisher criticizes mathematical statisticians for an over-emphasis on randomization tests because of their lack of parametric assumptions. He says in regard to the above test,

> The reader will realize that it was in no sense put forward to supersede the common and expeditious tests based on the Gaussian theory of errors. The utility of such nonparametric tests consists in their being able to supply confirmation whenever, rightly or, more often, wrongly, it is suspected that the simpler tests have been appreciably injured by departures from normality. [DOE, p. 48].

He goes on to say that experimenters should, when possible, rely on their knowledge of the variability of the experimental material. From his previous development of the purposes of randomization, his message seems clear. Nonetheless, there has been confusion concerning the role of randomization in normal theory tests. In his Fisher Memorial Lecture, Oscar Kempthorne (1966) argues that Fisher viewed the use of the normal theory distributions as valid only when they are adequate approximations to the exact distributions under the randomization structure. This, I think, is the correct reading of Fisher. Fisher's criticism simply reflects his realization that in many situations the normal theory distributions are adequate approximations.

3. The Initial Fisher-Neyman Controversy

It is of interest to recall Fisher's first strong criticism of Jerzy Neyman, since it involved an important issue in the analysis of randomized experiments. Fisher was the lead discussant of Neyman's 1935 paper, "Statistical Problems in Agricultural Experimentation", read before the Royal Statistical Society. Although he was unhappy with several aspects of the paper, Fisher focusses primarily on Neyman's analysis of the z-test (z is one-half the natural log of the F-statistic) for treatment effects. Neyman formulated a quite general model for treatment responses in randomized block and Latin square designs. Subject to equal variance of plot errors for all treatments and equal correlation of plot error for all pairs of treatments, his model allowed each treatment to respond in a different manner on each plot. In particular, Neyman made no assumption of treatments having fixed and additive responses on all plots. The null hypothesis he considered was that the average treatment response over the entire experimental area was the same for all treatments. Under this null hypothesis he found that the z-test for the randomized block design was unbiased in the sense that the expectation of the mean square for treatments was equal to the expectation of the mean square for error taken over all possible randomizations. But the test for the Latin square design was, in general, not unbiased with the expectation of the mean square for treatments being larger than the expectation of the mean square for error. With the additional assumption that the correlation of plot errors is one for all pairs of treatments, the z-test is unbiased.

Fisher chastised Neyman for obtaining an incorrect result for the Latin square and suggested that he was misled by his excessive use of symbolism; however, Fisher had entirely overlooked Neyman's null hypothesis. The null hypothesis Fisher was considering was that on any one plot all treatments have the same effect. In that case the correlation of plot errors is one for all pairs of treatments and Neyman's result correctly indicates that the z-test is unbiased.

Once Neyman had pointed out Fisher's oversight, Fisher maintained that the z-test was only intended for testing the null hypothesis of identical treatment effects on each plot. Neyman replied that he was merely examining the z-test under the assumptions which he considered most relevant to agriculture experimentation.

The assumption that treatments have constant and additive effects is crucial to the analysis of most randomized experiments. But, as Neyman observes, it is most likely violated to some degree. It is to Neyman's credit that he considered some of the resulting effects on the standard analysis.

4. Unwanted Random Assignments

A philosophical problem of randomization theory is the determination of the correct procedure in the event of an unwanted assignment of treatments to experimental units. An assignment is usually unwanted when it has a discernible pattern that, one suspects from experience, could yield poor estimates. The pattern might confound treatments with influential environmental factors producing poor estimates of treatment effect and, most likely, an underestimate of error. Alternatively, the pattern might balance treatments and environmental factors producing good point estimates of treatment effects at the expense of overestimating the error. A widely recommended approach has been to discard the assignment and draw a new assignment, but this changes the distribution of possible arrangements and consequently, to some degree, invalidates distributional results derived from the original randomization plan. A solution to this problem is to eliminate unwanted assignments from the outset. This is usually an extremely difficult task, and the approximate distributional properties under the original randomization scheme, such as those of the t-test in the paired comparison experiment, might well be inadequate under the new plan. If so, one must work with the actual distribution of possible assignments which could also be quite difficult.

It appears that Fisher never seriously addressed the problem, but there are indications that he favored some manner of validly eliminating undesirable assignments. He argued strongly against systematic, balanced designs and against adjusting randomizations to balance the design:

> Most experimenters on carrying out a random assignment of plots will be shocked to find how far from equally the plots distribute themselves; three or four plots of the same variety, for instance, may fall together at the corners where four blocks meet. This feeling affords some measure of the extent to which estimates of error are vitiated by systematic regular arrangements, for, as we have seen, if the experimenter rejects the arrangements arrived at by chance as altogether "too bad", or in other ways "cooks" the arrangement to suit his preconceived ideas, he will either (and most probably) increase the standard error as estimated from the yields; or, if his luck or his judgment is bad, he will increase the real errors while diminishing his estimate of them. [CP 48, pp. 90-91].

It seems reasonable to assume that Fisher would have felt the same, say, about systematic, regular arrangements when they were the result of a random selection, at least in the common randomized designs. In fact, L.J. Savage reports the following conversation with Fisher in 1952:

> "What would you do", I had asked, "if, drawing a Latin square at random for an experiment, you happened to draw a Knut Vik square?" Sir Ronald said he thought he would draw again and that, ideally, a theory explicitly excluding regular squares should be developed. (Savage, 1962, p. 88).

In any case, it is safe to say Fisher did not view the problem of undesirable assignments as an inherent flaw in the theory of randomized experiments.

References

Kempthorne, Oscar (1966). "Some Aspects of Experimental Inference," Journal of the American Statistical Association, 61, 11-34.

Neyman, J., K. Iwaszkiewicz, and St. Kolodziejczyk (1935). "Statistical Problems in Agricultural Experimentation," Supplement to the Journal of the Royal Statistical Society, 2, 107-180.

Savage, L.J. et al. (1962). The Foundations of Statistical Inference: A Discussion New York: John Wiley and Sons.

RANDOMIZATION AND DESIGN: II

Rick Picard

1. Introduction

Fisher contributed greatly to the advancement of experimental design. Prior to his work, little had been accomplished in the area. Before the 1920s, many people had conducted agricultural field trials, but there were no widely accepted techniques concerning their layout or analysis. Consequently, things were done in whatever manner pleased the experimenter, and statistical analyses of the time were crude and lacking in theoretical justification.

Topics that received the attention of experimenters included the determination of the size, shape, and number of plots that were best used in field trials, and the methods that cut time spent on computations. See Hall and Mercer (1911) as well as Wood and Stratton (1910) for work representative of the era.

Another subject of concern was the "correcting for position". Field experimenters had long realized that because some plots in their fields were superior to others there could result a favoring of whatever treatments may have been used on these plots. Several methods were proposed to deal with this problem, and two of them are illustrated here.

The first method was referred to as the contingency method (Pearl and Surface, 1916). Let x_{ij} denote the observed yield of the (i,j)th plot in a rectangular field. The advantage of the (i,j)th plot is computed by the formula $\frac{x_{i+}x_{+j}}{x_{++}} - \frac{x_{++}}{IJ}$, where the use of "+" as a subscript denotes summation over the corresponding subscript index, and I and J are the total numbers of rows and columns. The "corrected value" for the yield, as Pearl and Surface reasoned, should be

$$c_{ij} = x_{ij} - \left(\frac{x_{i+}x_{+j}}{x_{++}} - \frac{x_{++}}{IJ}\right).$$

By examining some data, they concluded that this method would reduce the correlation of adjacent plots and yield smaller estimates of error.

The second method was a proportional variant of the first. It entailed expressing the advantage of a given plot (as computed above) as a percentage of the average yield. The observed value would then be adjusted by the corresponding percentage. For example, if $x_{ij} = 50$, $x_{++}/IJ = 40$, advantage = 4, then this method would yield

$$c_{ij} = 50 - (4/40)50 = 45.$$

It is quite clear that the variety of methods proposed was due to the lack of established techniques, and reflected the crude but common-sense approaches toward recognized problems.

2. Early Contributions by Fisher

Fisher arrived at Rothamsted in 1919. His first paper concerning field experiments appeared two years later in the Journal of Agricultural Science [CP 15]. It was the first of several "Studies in Crop Variation" and dealt primarily with the history of the Broadbalk wheat fields over the previous 70 years. It was non-statistical and no discussion of randomization or design was included. It was in his 1923 paper, "Studies in Crop Variation II" [CP 32], written with W. Mackenzie, that the groundwork was laid (see also Cochran's discussion of this paper in this volume). It marked the first published application of the analysis of variance, dealing with the effects of dung and potash on potato yields at Rothamsted. Even early in 1923, it appeared that Fisher might have had his first inkling of the uses of randomization, since in the 1923 paper he and Mackenzie partitioned the total sum of squares into the between the within sum of squares and made the following statement:

> Furthermore, if all the plots were undifferentiated, as if the numbers had been mixed up and written down in random order, the average value of each of the two parts is proportional to the number of degrees of freedom in the variation of which it is compared.

Given this statement, it is not surprising that, in his second major contribution to design, the 1925 first edition of Statistical Methods for Research Workers

[SMRW], Fisher introduced the idea of randomization.[1] In fact, his position on the subject was quite firm. "Systematic arrangements in plots should be avoided," he said, and he gave reasons for such a position. Also, some examples of both the randomized block design and the Latin square design were worked out, using the Hall and Mercer data from Rothamsted.

Another feature of Fisher's 1925 book was the great emphasis on significance tests, with many tables of distributions to aid in their use. The emphasis was a result of Fisher's general outlook, later expressed in the Design of Experiments [DOE]: "Every experiment may be said to exist only in order to give the facts a chance of disproving the null hypothesis". This outlook was a source of the criticism that some had of the book, but Fisher was quite persistent in this respect. The book was subject to another criticism, which Fisher anticipated at the beginning -- namely the lack of mathematical proofs. Since Fisher intended the book to be a useful tool for biologists and other research workers in analyzing data they would encounter, as he pointed out in the preface of the 2nd edition in 1928, he decided that the inclusion of proofs was simply unnecessary.

In spite of all the criticism it provoked, the first edition of Statistical Methods for Research Workers [SMRW] was a milestone. Many important experimental techniques not previously alluded to by Fisher -- randomized blocks, Latin squares, split plots, and uses of randomization -- surfaced at once in its pages.

3. Further Contributions

Fisher advanced several new ideas on design in his paper "The Arrangement of Field Experiments" [CP 48], published in 1926. This was written partly in response to a paper earlier that year by the Rothamsted director, Sir John Russell (1926),

---

[1] It is worth mentioning here that Neyman wrote a paper concerning the use of randomization in what we now know as a completely randomized design in this same year. See Neyman (1923) as well as a footnote by Scheffé (1959, p. 291). Twelve years later, when Neyman extended these randomization models to randomized blocks, a heated dispute between he and Fisher arose concerning them. See Holschuh's presentation in this volume for further details.

who advocated the "vary one factor at a time" approach to experimentation. Fisher expounded the value of more complex experiments and introduced the idea of factorial design. In this work, he mentioned particularly "Student's" (1923) paper and systematic design -- a topic that would mushroom into a controversy later. He also mentioned the possibility of re-randomizing in the event of a bad first draw, and here he was clearly against it. Fisher's first mention of confounding appeared in the last paragraph of the paper, where he commented on the rationale behind confounding, as well as the term itself. The first detailed account of the subject appeared three years later, in the paper "Studies in Crop Variation VI" with Eden [CP 78], where an example was given. This paper was somewhat confusing as to what was happening and why (some confounded interactions were pooled into error, and no explanation was made). Though Fisher was the one to initiate the idea of confounding, the development of the subject was primarily the work of others (Yates, Bose, etc.). Cochran's lecture in this volume also contains some interesting remarks on these papers.

Another idea that Fisher introduced in the early 1930s was that of analysis of covariance. In the 1932 edition of [SMRW], Fisher took data from a uniformity trial of tea bushes in Ceylon to show how this new technique could be used to reduce experimental error.

4. Randomized Versus Systematic Designs: Tedin's Experiment

Fisher's work in design prompted several disputes, mostly concerning his views on randomization and the comparison of randomized designs to systematic ones. A particular piece of work to which Fisher referred was by the Scandinavian Olaf Tedin.

In 1931, Tedin investigated the properties of systematic Latin squares as opposed to squares drawn at random. Tedin took results from eight uniformity trials conducted between 1915 and 1925. From these trials, he excerpted 91 5×5 squares of data. He then took twelve treatment arrangements -- two "Knight Move" squares, two "Diagonal" squares, seven squares chosen at random, and one square he viewed as "most compact". For each of these 12 assignments, he took his 91 squares of data and then, pretending that these were realizations of experiments, he calculated the analysis of variance tables and the resulting values of $V = SS_{trt} \div (SS_{trt} + SS_{err})$ -- essentially the F values. He gave results on "Knight Move" and "Diagonal" squares in the following table:[2]

Table 1: Results of "Knight Move" and "Diagonal" Squares

Normal-theory Deciles of V	0.000	0.079	0.112	0.156	0.192	0.229	0.269	0.314	0.371	0.453	1.000	Expected No. per Class	χ^2	P
Knight Move #1		16	12	7	10	7	2	11	12	7	7	9.1	15.04	0.09
Knight Move #2		7	13	7	20	9	6	9	4	10	6	9.1	19.48	0.02
Knight Move Total		23	25	14	30	16	8	20	16	17	13	18.2	20.42	0.016
Diagonal #1		10	7	10	4	12	8	9	10	11	10	9.1	5.15	0.82
Diagonal #2		9	5	6	15	8	6	6	13	5	18	9.1	20.05	0.02
Diagonal Total		19	12	16	19	20	14	15	23	16	28	18.2	10.97	0.28

Comparing his results (from Table 1 above) to the theoretical normal values provided by Fisher, Tedin discovered that his "Knight Move" squares significantly ($P = 0.01$) underestimated them, that the "Diagonal" squares overestimated them, but that the random squares were consistent with normal theory results. Thus he concluded:

> The practical field experimenter who desires the highest degree of scientific accuracy in his experiments, and especially in his estimate of the validity of eventual conclusions, should therefore be careful to avoid systematic plans and must choose his arrangements at random.

This was one of the results that Fisher liked to quote in his later discussions of randomization, particularly with "Student".

---

[2] Tedin did not detail in a similar fashion the results for his random squares.

5. Randomized Versus Systematic Designs: Fisher and "Student"

The argument between Fisher and "Student" over their conflicting ideas on design resulted in an ongoing debate. Fisher was a confirmed believer in randomization while "Student" had gone on record as saying that he would not use any Latin square other than a Knight Move (see in particular, p. 401 of Pearson and Kendall, 1970). The seeds of the argument were planted early in 1923, and their bickering over the subject blossomed for years. But the battle was joined in 1936 when "Student" presented a paper before the Royal Statistical Society entitled "Co-operation in Large Scale Experiments" (1936). At one point in the paper, "Student" said: "A balanced arrangement like the Half Strip Drill is best if otherwise convenient". The "Half Strip Drill Method" or "Sandwich Design" had been proposed by Beavin for avoiding the effect of fertility trends in comparative field experiments involving two treatments. The design was laid out as follows:

treatment A	
treatment B	
treatment B	
treatment A	↑ fertility trend
treatment A	
treatment B	
treatment B	
treatment A	

Beavin reasoned that neither treatment would be favored by this arrangement and that an "accurate" estimate of the treatment difference would be obtained.

Fisher, as one of the discussants of the paper, responded with his customary arguments: (i) No valid estimate of error is obtained from a balanced arrangement, and how to compute any estimate is unclear. (ii) Though "Student" claimed "actual errors" are smaller in balanced designs, such errors are always unknown. (iii)

Tests of significance are vitiated.[3] (iv) Systematic misapplication of treatment would ruin the experiment.

"Student's" reply to these arguments was brief and cordial -- "This is an old matter of controversy between Professor Fisher and myself. He says to me 'your half drill strips have no validity and conclusions cannot be drawn from them'. I say to him 'your errors are so large that no conclusions are drawn'. Neither of these criticisms is true, and one is about as good as the other."

To answer "Student's" reply, Fisher and Barbacki [CP 139] took data from a uniformity trial on wheat conducted by Gustav Wiebe, and analyzed it to compare the half drill strip to randomized designs. Fisher extracted from the trial a 16×12 rectangular array of plots, and from this he created 48 sandwiches as illustrated below:

	1	2	3	4	5	6	7	8	9	10	11	12
1	A	A	A	A	A	A	A	A	A	A	A	A
2	B	B	B	B	B	B	B	B	B	B	B	B
3	B	B	B	B	B	B	B	B	B	B	B	B
4	A	A	A	A	A	A	A	A	A	A	A	A
5	A	A	A	A	A	A	A	A	A	A	A	A
6	B	B	B	B	B	B	B	B	B	B	B	B
7	B	B	B	B	B	B	B	B	B	B	B	B
8	A	A	A	A	A	A	A	A	A	A	A	A
9	A	A	A	A	A	A	A	A	A	A	A	A
10	B	B	B	B	B	B	B	B	B	B	B	B
11	B	B	B	B	B	B	B	B	B	B	B	B
12	A	A	A	A	A	A	A	A	A	A	A	A
13	A	A	A	A	A	A	A	A	A	A	A	A
14	B	B	B	B	B	B	B	B	B	B	B	B
15	B	B	B	B	B	B	B	B	B	B	B	B
16	A	A	A	A	A	A	A	A	A	A	A	A

[3] Despite this comment, it is interesting that on occasion Fisher was not above analyzing systematic experiments as though they were randomized. One example in Statistical Methods for Research Workers [SMRW] is the Rothamsted potato data analyzed as a split plot.

Four analyses were carried out: (a) Each sandwich was used to provide an estimate of the treatment difference. These 48 numbers were then used to compute $\widehat{SE}_{A-B} = 2218.50$. (b) "Randomized Sandwiches" were used, where the treatment assignment within each sandwich was randomly chosen between

$$\begin{bmatrix} A \\ B \\ B \\ A \end{bmatrix} \text{ or } \begin{bmatrix} B \\ A \\ A \\ B \end{bmatrix}$$

Result: $\widehat{SE}_{A-B} = 2353.35$.

(c) The procedures (a) and (b) were repeated with each sandwich viewed as two pairs. The corresponding $\widehat{SE}$'s for (a) and (b) were 3020.35 and 3063.84. Fisher and Barbacki concluded:

> It will be noticed that not only has the systematic experiment the higher real error, but it yields a lower estimate of error than the randomized experiment. The test of significance is vitiated for both reasons.
>
> It should be noticed that when a systematic experiment has been carried out, there is no more reason for estimating the error from pairs than from sandwiches ... neither has, in fact, any objective justification -- when randomization been practiced, there is no such ambiguity.

"Student's" reply to Fisher and Barbacki showed up immediately in a letter to Nature (1936, p. 971) in which his main objections were as follows. First, it is quite inappropriate to treat the 48 sandwich values as independent, as did Fisher in his analysis. Second, Fisher's first conclusion ran counter to the work of Tedin and others. In any event, basing any broad conclusions on one analysis of a single data set was certainly unjustified. "Student" also summarized what he felt was a more appropriate analysis and gave error estimates. Fisher answered this immediately[4] in a letter [CP 140], saying "I cannot imagine that Student's new estimate is any less arbitrary than the others".

---

[4]As was the case in 1926 when he differed with Sir John Russell's views on experimentation, here too Fisher replied as quickly as possible and in the same publication -- he wrote a letter to Nature.

"Student" was drafting a more detailed reply to this when he died in 1937. In the posthumously published response ("Student", 1937), he expanded on his letter to Nature, repeating his old criticisms and adding a new one -- the dilemma of what to do in the event that a "bad" randomization is drawn (a criticism to which Fisher now avoided response). As for Fisher's criticism on the ambiguity in computing error estimates, "Student" suggested using any suitable estimate -- chosen independently of the data, of course. "Student" also proposed a new method of arranging treatments in field experiments. Taking the Fisher and Barbacki layout as an example, he suggested "sandwiches in both directions," as:

```
A B B A | A B B A
B A A B | B A A B    . . .
B A A B | B A A B
A B B A | A B B A
--------+--------
A B B A | .
B A A B |   .
B A A B |     .
A B B A |
   .
   .
   .
```

Performing an analysis for this arrangement, "Student" concluded that the size of the error estimate was comparable to random sandwiches, an arrangement he viewed as Fisher's concession to balancing.

This final paper by "Student" was published in Biometrika through the efforts of Neyman and E.S. Pearson, who added a postscript saying that "Student" had misperformed his analysis, and offered a substitute -- reaching the same conclusions all over again.

6. Randomized Versus Systematic Design: E.S. Pearson and Yates

A year later, E.S. Pearson (1938) wrote a paper which stated his view on the matter debated by Fisher and "Student". Pearson had worked a great deal on this subject with "Student", and was able to clarify some of the points that were muddled by the rhetoric of Fisher and "Student".

He analyzed data from uniformity trials and noted the conservative nature of the F test under the null hypothesis. From his analysis, he also concluded in the cases he dealt with that balanced designs found large treatment differences somewhat more often than did randomized designs. To show why, he considered a randomized block design. He let $x_{i(r)_s}$ denote the response of plot r in block i when it was assigned treatment s. The model considered was

$$x_{i(r)_s} = m_{i(r)} + \delta_s,$$

where $m_{i(r)}$ was the fixed plot response and δ_s was the additive treatment effect. It was assumed that $\Sigma\delta_s = 0$. From the standard analysis of variance,

$$SS_{\text{treatments}} = (\#\text{ blocks}) \sum_s (\text{treatment 's' mean} - \text{grand mean})^2$$

$$= B \sum_s (m_{\cdot(r_s)} + \delta_s - m_{\cdot(\cdot)})^2,$$

where $m_{\cdot(r_s)}$ is the mean of fixed plot responses of plots assigned treatment s. Expanding the above

$$SS_{\text{treatments}} = B\{\sum_s (m_{\cdot(r_s)} - m_{\cdot(\cdot)})^2 + \sum_s \delta_s^2 + 2\sum_s \delta_s (m_{\cdot(r_s)} - m_{\cdot(\cdot)})\}$$

$$= B\{S_1 + S_2 + S_3\}.$$

We also have from the standard analysis

$$SS_{\text{error}} = \sum_i \sum_s (x_{i(r)_s} - x_{i(\cdot)} - x_{\cdot(r_s)} + x_{\cdot(\cdot)})^2$$

$$= \sum_i \sum_s (m_{i(r)_s} - m_{i(\cdot)} - m_{\cdot(r_s)} + m_{\cdot(\cdot)})^2.$$

The F test for treatments is

$$F = (\text{constant}) \frac{S_1 + S_2 + S_3}{SS_{\text{error}}}.$$

If the experiment were "perfectly balanced", that is, the assignments of treatments to units was such that all treatments had the same average fixed plot response ($m_{\cdot(r_s)} = m_{\cdot(\cdot)}$ $\forall s$), the F value above would reduce to

$$F = (\text{constant}) \frac{\sum_s \delta_s^2}{\sum_i \sum_r (m_{i(r)} - m_{\cdot(\cdot)})^2} .$$

Comparing this F value to a tabled F distribution, it is found that the experiment <u>never</u> rejects the hypothesis of no treatment effects for small $\sum_s \delta_s^2$, but <u>always</u> rejects for large values. The resulting power curve, a step function, is very different to the one based on random assignment of treatments to unit. Pearson compared the two curves over regions for large $\sum_s \delta_s^2$ and decided the balanced assignment was superior. This result was empirically verified by use of data from uniformity trials where hypothetical treatment effects were added. Plotted results were quite similar to the following:

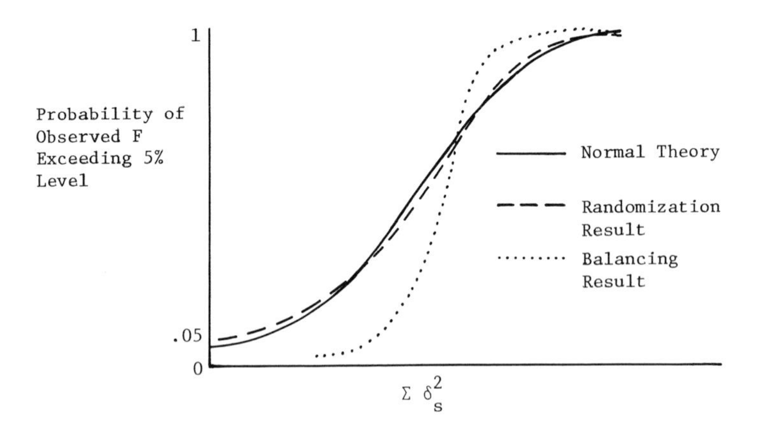

Pearson granted that his results gave credence to one of Fisher's contentions -- that the behavior of the computed F value for balanced designs was quite unlike behavior under the standard assumptions -- but concluded:

> Nevertheless, it might be argued with reason that useful conclusions for the practical agriculturalist regarding treatment differences cannot be drawn until they can be based on something approaching certainty; in this region the balanced layout seemed to give a slight advantage.

Yates joined the debate in 1939 (from this point on, Fisher added little). He noted (Yates, 1939) that in a large field there are a variety of randomized designs (e.g., side-by-side Latin squares) not previously considered that maintain some of the balance desired by "Student" and yet yield valid tests of significance. Yates proposed one such design and then closed with a summary of arguments that favor the use of randomization in designed experiments.

While it is clear that many questions concerning the relative merits of systematic design as opposed to randomization have yet to be resolved, the "Student"-Fisher dispute on half drill strips did much to clarify the issues involved.

7. Conclusion

It seems appropriate to close with a quote from Savage (1976):

> Fisher is the undisputed creator of the modern field that statisticians call the design of experiments, both in the broad sense of keeping statistical considerations in mind in the planning of experiments and in the narrow sense of exploiting combinatorial patterns in the layout of experiments.

References

Hall, A.D. and W.A. Mercer (1911). "The Experimental Error of Field Trials," Journal of Agricultural Science, 4, 107-132.

Neyman, J. (1923). "Sur les Applications de la Théorie des Probabilitiés aux Expériences Agricoles: Essay des Principes," Roczniki Nauk Rolniczch, 10, 1-51.

Neyman, J. and E.S. Pearson (1937). "Note on Some Points in Student's Paper on 'Comparison Between Balanced and Random Arrangements of Field Plots'," Biometrika, 30, 380-388.

Pearl, R. and F.M. Surface (1916). "A Method of Correcting for Soil Heterogeneity in Variety Tests," Journal of Agricultural Research, 5, 1039-1050.

Pearson, E.S. (1938). "Some Aspects of the Problem of Randomization. II. An Illustration of 'Student's' Inquiry into the Effect of Balancing in Agricultural Experiments," Biometrika, 30, 159-171.

Pearson, E.S. and M.G. Kendall (1970). Studies in the History of Statistics and Probability. Volume 1. London: C. Griffin and Company.

Russell, J. (1926). "Field Experiments: How They Are Made and What They Are," Journal of the Ministry of Agriculture, 32, 989-1001.

Savage, I.R. (1976). "On Re-Reading R.A. Fisher," The Annals of Mathematical Statistics, 4, 442-475.

Scheffé, H. (1959). The Analysis of Variance. New York: John Wiley and Sons.

"Student" (1923). "On Testing Varieties of Cereals," Biometrika, 15, 271-293.

"Student" (1936). "Co-operation in Large Scale Experiments," Supplement to the Journal of the Royal Statistical Society, 4, 115-136.

"Student" (1937). "Comparison Between Balanced and Random Arrangements of Field Plots," Biometrika, 29, 363-379.

Tedin, O. (1931). "The Influence of Systematic Plot Arrangements Upon the Estimate of Error in Field Experiments," Journal of Agricultural Science, 21, 191-208.

Wood, T.B. and F.J.M. Stratton (1910). "The Interpretation of Experimental Results," Journal of Agricultural Science, 3, 417-440.

Yates, F. (1939). "The Comparative Advantages of Systematic and Randomized Arrangements of Field Plots," Biometrika, 30, 441-464.

BASIC THEORY OF THE 1922 MATHEMATICAL STATISTICS PAPER

Seymour Geisser

1. Introduction

Some of Fisher's papers prior to 1922 had contained brief glimpses of his general ideas on statistical theory. The first full account of these ideas appeared in 1922 in the 59-page article "On the Mathematical Foundations of Theoretical Statistics" [CP 18]. The paper opens with general remarks about the then-current state of theoretical statistics.

According to Fisher, there are two reasons for the neglect of theoretical statistics: (i) a philosophical reason -- since statistics is the study of results subject to greater or lesser error, it is thought that precision of the concepts or ideas is either impossible or not a practical necessity, and (ii) a technical reason -- the confusion that exists between statistics and parameters.

Fisher believed that the purpose of statistical methods is to reduce a large quantity of data to a few quantities, which can adequately represent the whole in that it will contain as much as possible of the relevant information in the original data. This is accomplished by constructing a hypothetical infinite population specified by a relatively few parameters. Generally speaking, data will supply a large number of "facts", many more than are sought. That is to say, much information in the data is irrelevant. Statistical analysis via the reduction of the data is aimed at extracting the relevant information and thus excluding the irrelevant information.

The problems of statistics, as stated by Fisher, are (i) specification -- choice of the distribution law, (ii) estimation -- choice of a statistic for estimating the parameters, and (iii) sampling distributions for the statistics involved in (ii). The present lecture focusses on (ii) and (iii).

2. Consistency

A statistic is consistent (Fisher consistent) if, when calculated from the whole population, it is equal to the parameter describing the probability law. Thus for every n let $F_n(x)$ be the empirical distribution function and let $T_n = g(F_n(x))$ be an estimator of θ. Then T_n is Fisher consistent (F.C.) for θ if $g(F(x|\theta)) = \theta$.

Example 1. Let $X_1,\ldots,X_n$ be independent (i.i.d.) $N(\theta,\sigma^2)$. Define $\bar{X}_n$ to be $\frac{1}{n}\sum_1^n X_j$, then $\bar{X}_n$ is F.C. since $\int x\, dF(x|\theta) = \theta$. But $T_n = \bar{X}_n + \frac{1}{n}$ is not F.C. although it is consistent in the ordinary (weak convergence) sense.

The ordinary definition of consistency has its problems as pointed out by Fisher. Suppose that T_n is consistent in the ordinary sense and that for an arbitrary n_1, another estimate is defined by

$$T'_n = \begin{cases} A_{n_1} & \text{for } n \leq n_1 \\ \dfrac{n_1 A_{n_1} + (n-n_1)T_{n-n_1}}{n} & \text{for } n > n_1. \end{cases}$$

Then T'_n is also consistent regardless of the value A_{n_1} assumes. Fisher obviously did not like this kind of arbitrariness and made great sport of it.

Example 2. Let $X_1,\ldots,X_n$ be i.i.d. $N(\mu,\sigma^2)$ with σ^2 unknown. Consider two estimates of σ: $\hat{\sigma}_1 = (\sum_{i=1}^n |x_i-\bar{x}|^2/n)^{\frac{1}{2}}$, $\hat{\sigma}_2 = \frac{1}{n}\sqrt{\frac{\pi}{2}}\sum_{i=1}^n |x_i-\bar{x}|$. Since both estimators are F.C. we need to have a second criterion to differentiate between them to say which one is superior. One suggestion is to compare their variances. When a large sample and asymptotic normality are assumed, Var $(\hat{\sigma}_1) = \frac{\sigma^2}{2n}$ and Var $(\hat{\sigma}_2) = \frac{\sigma^2(\pi-2)}{2n}$, i.e., Var $(\hat{\sigma}_2) >$ Var $(\hat{\sigma}_1)$. Thus $\hat{\sigma}_1$ can be considered to be superior to $\hat{\sigma}_2$; Fisher interprets the asymptotic variance ratio of 1.14 in terms of equivalent sample sizes.

3. Efficiency

A statistic is efficient if its large sample distribution is normal with smallest possible variance (generally, the Fisher information reciprocal). Fisher recognizes the lack of uniqueness associated with the efficiency of statistics, and moves on to consider sufficiency.

4. Sufficiency

A statistic is sufficient if it summarizes the whole of the relevant information supplied by the sample. If θ is to be estimated and T_1 is sufficient, then for any other statistic T_2, we have that T_2 given T_1 is independent of θ.

Fisher "demonstrates" that sufficient statistics are efficient under the assumption of normality. He argues roughly as follows. Suppose T_1 and T_2 are asymptotically bivariate normal with mean θ, variances σ_1^2 and σ_2^2, and correlation coefficient r so that T_2 given T_1 is distributed as

$$N(\sigma + \frac{r\sigma_2}{\sigma_1} (t_1 - \theta), \sigma_2^2(1 - r^2)).$$

If T_1 is sufficient, then this conditional distribution should not involve θ. This implies that

$$E(T_2|T_1) = r \frac{\sigma_2}{\sigma_1} t_1. \tag{1}$$

Taking expectation on both sides of expression (1) with respect to T_1, we get

$$\theta = E(T_2) = E_{T_1} E(T_2|T_1) = r \frac{\sigma_2}{\sigma_1} \theta. \tag{2}$$

Thus $r\sigma_2 = \sigma_1$, which yields $\sigma_1 \leq \sigma_2$ since $r \leq 1$. This completes the "demonstration".

5. Fisher's Comment on Pearson's Method of Moments

Pearsonian curves have four parameters and are fitted by means of the first four moments; hence, they require at least four finite moments. Further, if the variance of the fourth moment decreases with n, then the eighth moment must also be finite. Distributions not possessing the requisite moments cannot be determined by Pearson's method. A standard example is that of the Cauchy distribution with center μ, where X has probability density function (p.d.f.)

$$f(x) = \frac{1}{\pi[1 + (x-\mu)^2]}.$$

The question is how to estimate μ from n independent observations. The first sample moment, $\bar{X}$, is useless since $\bar{X}$ has the same distribution as X. Yet this is the result we would get if the method of moments were used. The median provides a consistent estimate, which shows that the problem of estimating μ can be solved by "adopting adequate statistical methods".

Fisher points out that for data such as those coming from a Cauchy distribution "rejection of extreme values is often advocated ... unless there are other reasons for rejection than mere divergence from the majority, it would be more philosophical to accept these extreme values ... as indications that the distribution of errors is not normal". He argues, then, that data should be recorded in full, no matter what estimate is used.

6. Method of Maximum Likelihood

Since sufficiency is not adequate as a criterion to provide estimators in general, Fisher proposes maximum likelihood estimation (MLE) as a direct method of obtaining a sufficient estimate. The approach he takes is as follows. Let $\Pr[x \in dx] = f(x|\theta_1,\ldots,\theta_r)dx$, then for n independent observations,

$$\Pr[n_1 \text{ x's} \in dx_1,\ldots, n_p \text{ x's} \in dx_p] = \frac{n!}{\prod_{j=1}^{p} n_j!} \prod_{j=1}^{p} \{f(x_j|\theta_1,\ldots,\theta_r)dx_j\}^{n_j}. \tag{3}$$

Maximizing equation (3), or equivalently $\sum_{j=1}^{p} n_j \log f(x_j|\theta_1,\ldots,\theta_r)dx_j$, with respect to $\theta_1,\ldots,\theta_r$ yields estimates for $\theta_1,\ldots,\theta_r$.

In his earlier argument for maximum likelihood estimation [CP 1], Fisher had taken a Bayesian approach because the maximizing procedure resembles the calculation of the mode of a posterior probability. In the present paper he is very concerned to differentiate MLE from the Bayesian approach. He argues against the "customary" Bayesian use of flat priors on the grounds that different results obtained when different scales for the parameters are considered.

To illustrate Fisher's argument, suppose x denotes the number of successes out of n independent (i.i.d.) trials with probability of success p; then the likelihood function is $L = \frac{n!}{x!(n-x)!} p^x(1-p)^{n-x}$ which is maximized when p is chosen to be x/n. Now, if a uniform distribution on (0,1) is taken to be the prior distribution of p, then the Bayesian analysis would yield $\Pr[p \in dp] \propto p^x(1-p)^{n-x}dp$. But, if we parameterize this Bernoulli process in a different way, say in terms of θ with $\sin\theta = 2p-1$, then the likelihood function of θ is

$$L_\theta = \frac{n!}{x!(n-x)!} \frac{(1+\sin\theta)^x}{2^x} \frac{(1-\sin\theta)^{n-x}}{2^{n-x}},$$

which, when maximized with respect to θ, gives $\sin\hat{\theta} = \frac{2x-n}{n} = 2\hat{p}-1$. Thus the MLE is invariant under 1-1 transformation. What about a Bayes estimate? What prior should be assigned to θ? Uniformity of θ on $(-\frac{\pi}{2},\frac{\pi}{2})$ leads to a posterior distribution of p proportional to $p^{x-\frac{1}{2}}(1-p)^{n-x-\frac{1}{2}}$, which is different from the previous result $p^x(1-p)^{n-x}$. Due to this inconsistency, Fisher derides the arbitrariness of the Bayes prior and chooses not to adopt the Bayesian approach. (Jeffreys later proposes the square root of Fisher's information measure as a prior on p, which leads to transformation-invariant results, because the Fisher information, I_θ, satisfies the relation $I_\theta = I_p(\frac{dp}{d\theta})^2$.)

Fisher suggests using L_p as a relative measure of the plausibility of various values of p and introduces the term "likelihood" to distinguish the concept from probability, confessing that earlier he had confused the two. He says that "likelihood is not here used loosely as a synonym for probability, but simply to express the relative frequencies with which such values of the hypothetical quantity p would in fact yield the observed sample".

7. Probable Error of the Maximum Likelihood Estimate

Assuming that the distribution of the MLE tends to normality, Fisher demonstrates that the variance is the reciprocal of the Fisher information. That is, if T is an optimal statistic satisfying $\partial L/\partial\theta = 0$ at $\theta = T$, then the variance of the large-sample normal distribution of T is the inverse of the mean of $-\partial^2 L/\partial\theta^2$. The general consistency of the MLE is not used, and the "proof" relies heavily on the initial assumption.

Fisher stresses that the "proof applies only to statistics obtained by the method of maximum likelihood", and notes the earlier (pre-1914) confusion on this point by K. Pearson and others, who had used the same variance result for method of moments estimates. He points out "the very remarkable circumstance" that when (in

1914) the formula was found to be incorrect for method of moments estimates, the inefficiency of this method was not revealed.

8. Sufficiency of Maximum Likelihood

Fisher has now derived the explicit form of the limiting normal distribution for the MLE, prior to which (§ 4) he has demonstrated that a sufficient estimate has the smallest-variance normal distribution in large samples. The theory would be complete if the MLE were found to be sufficient, since then the reciprocal of Fisher information would be a lower bound against which efficiency could be measured. Fisher claims to show that the MLE is sufficient, although his wording is ambiguous in some passages. His incorrect demonstration begins with the joint distribution of the MLE $\hat{\theta}$ and an arbitrary statistic T, whose joint density f satisfies $\partial \log f(\hat{\theta},T;\theta)/\partial\theta = 0$ at $\theta = \hat{\theta}$, according to Fisher. The factorization

$$f(\hat{\theta},T;\theta) = g(\hat{\theta};\theta)h(\hat{\theta},T)$$

is then deduced, and the sufficiency of $\hat{\theta}$ follows.

Taking the reciprocal of $E(-\partial^2 L/\partial\theta^2)$ as a lower bound on variance, Fisher illustrates the calculations on the Pearson Type III error curve with p.d.f.

$$f(x;\mu,a,p) \propto \left(\frac{x-\mu}{a}\right)^p \exp\{-(x-\mu)/a\},$$

where only μ is unknown and to be estimated. The method of moments estimator is $\mu = \bar{X} - a(p+1)$ with variance $a^2(p+1)/n$, while the asymptotic variance of the MLE is

$$[E(-\partial^2 L/\partial\mu^2)]^{-1} = a^2(p-1)/n.$$

Thus the method of moments can have zero efficiency.

9. Other Topics

The remainder of the paper contains mainly applications of maximum likelihood techniques and various efficiency calculations. There is a long discussion of the Pearson system of frequency curves. The final part of the paper looks at discrete distributions, where the method of minimum chi-square is related to maximum likelihood and the problem of grouped normal data is addressed in detail. The final example deals with the distribution of observations in a dilution series, and it is worthy of a careful examination.

After earlier displaying the potential lack of efficiency inherent in uncritical application of the method of moments, Fisher in an ingenious volte-face produces an estimation procedure for a dilution series example, which, though inefficient, is preferable to a fully efficient one essentially for economic and practical reasons. To be sure, in later years Fisher fulminated against the wholesale introduction of utility or decision theory into scientific work, but rarely again were such principles so elegantly and unobtrusively applied to such a significant practical problem.

An important monitoring procedure, of ongoing interest and wide applicability, used in this instance for estimating the density of protozoa in soils, was brought to Fisher's attention. A series of dilutions of a soil sample solution was made such that each is reduced by a factor a. At each dilution a uniform amount of the solution is deposited on s different plates containing a nutrient. After a proper incubation period, the number of protozoa on each plate is to be counted. A reasonable model for such a situation is that the chance of z protozoa on a plate is Poisson distributed with expected value θ/a^x, where θ is the density or number of protozoa per unit volume of the original solution, and x is the dilution level. A large number of such series were made daily for a variety of organisms. It proved either physically impossible or economically prohibitive to count the number of such organisms on every plate for many such series in order to estimate θ. First, Fisher suggests that only those plates containing no organisms be counted, so that the chance of such an occurrence at level x is $p_x = \exp(-\theta/a^x)$. Hence an experimentally feasible situation is attained which produces a joint likelihood for Y_x, the number of sterile plates at level x, as

$$L = \prod_{x=0}^{k} \binom{s}{y_x} p_x^{y_x} (1-p_x)^{s-y_x}.$$

He then calculates the contribution of a plate at level x to the information about $\log\theta$ to be $w_x = p_x(1-p_x)^{-1}(\log p_x)^2$; this is informative as to the number of dilution levels necessary in such experiments. Further, the total expected information is approximately given as

$$s \sum_{x} w_x \approx s\pi^2/(6 \log a).$$

The maximum likelihood solution to this problem, however, required a heavy investment of time and effort given the computational facilities of 1922.

At this point, Fisher demonstrates a second twist that helps make the problem tractable. He suggests that the expected total number of sterile plates be equated to the observed total number in order to obtain an estimate of θ. This "rough" procedure has expected information with respect to $\log \theta$ of approximate value

$$\frac{s}{\log 2 \log a}.$$

This results in a very quick and easy procedure possessing an efficiency, independent of the dilution factor, of about 88 percent.

This research was supported by National Institutes of Health Grant GN25-271 to the University of Minnesota.

FISHER AND THE METHOD OF MOMENTS

Roy Mensch

1. Introduction

The introduction of the method of maximum likelihood by R.A. Fisher offered statisticians a procedure for parameter estimation and some related optimality criteria. The principal tool that had been available previously was the method of moments, developed by Karl Pearson some twenty years earlier. Fisher directly attacked the method of moments in much of his work, and attempted to show the superiority of his likelihood techniques. While Fisher's views prevailed, the fact that the method of maximum likelihood (ML) and the method of moments (MM) were concerned, in part, with different questions belies any claim to actual superiority of one over the other. This lecture presents a comparison of the two methods and the goals of each. In it we review and summarize Pearson's work and discuss some of Fisher's criticisms. Many of the other presentations in this volume contain related discussions.

E.S. Pearson (1938) wrote of his father's early work: "In the '90's the biometric investigations into evolution and heredity had at times been held up because the development of statistical theory was unable to keep pace with the demands made on it...". Since the fundamental concern of statistics, in Karl Pearson's eyes, was "to predict from the past what will happen in the future", what was needed was a method for translating observed data into a predictive model. His solution was to develop a classification system of curves that fit observed biological data closely. We describe the system briefly in the next section.

Fisher's use of curve-fitting differed from that of Pearson, in the sense that Fisher assumed that he had already obtained a model. He then used the ML method to estimate the value of the parameters in the model. The superiority of the ML estimation procedure with regard to the proposed criteria of consistency, sufficiency, and efficiency rests on the initial assumption that the model is correct.

2. Pearson's Approach

We begin by outlining Pearson's approach to curve selection. With observed data, Pearson advocated the selection of a distribution, and then the use of the method of moments to estimate the values of parameters in the distribution. Finally he would carry out a test of goodness-of-fit for the distribution chosen. If the fit were not satisfactory, he would pick another distribution and start again. Pearson considered, as candidates for the choice of distribution, unimodal probability density curves which start at 0, rise to a maximum and then fall to zero again, i.e., for the density, y, that dy/dx = 0 at the maximum and at the ends of the range of x. Under this assumption, which appears to be reasonable for the biological data which Pearson was primarily interested in, a rather general model for the density is given by

$$\frac{dy}{dx} = \frac{y(x+a)}{F(x)}, \tag{1}$$

for some function F. Expansion of F(x) from expression (1) in polynomial form yields

$$\frac{dy}{dx} = \frac{y(x+a)}{b_0 + b_1 x + b_2 x^2 + \ldots} \doteq \frac{y(x+a)}{b_0 + b_1 x + b_2 x^2}. \tag{2}$$

For the approximate version of expression (2), only three parameters b_0, b_1, and b_2 are to be determined.

The selection of values for b_0, b_1, and b_2 generates the Pearsonian curve system. An itemization of this system, as listed by Elderton (1953), is given in Table 1 on the next page.

For example, let $b_1 = b_2 = 0$ in the approximate expression of (2) above. Then

$$\ln y = \int \frac{x+a}{b_0}\, dx = \frac{(x+a)^{\frac{1}{2}}}{2b_0} + K,$$

yielding $y = K\{\exp(x+a)^2/2b_0\}$, the form of a normal distribution.

As another example, let the roots of $b_0 + b_1 x + b_2 x^2$ be real and equal. Then

Table 1: The Pearsonian Curve System

Type	Equation for Density y
I (Beta)	$(1 + \frac{x}{a_1})^{m_1} (1 - \frac{x}{a_2})^{m_2}$
IV	$[1 + (\frac{x}{a})^2]^{-r+2/2} \exp\{-v \tan^{-1} (\frac{x}{a})\}$
VI (F-distribution)	$(x-a)^{m_2} x^{-m_1}$
Transition Normal	$e^{-\frac{1}{2}x^2}$
II (I)	$\{1 - (\frac{x}{a})^2\}^m$
VII (IV) (t-distribution)	$(1 + \frac{x^2}{a^2})^{-m}$
III (gamma)	$(1 + \frac{x}{a})^m e^{-\gamma x}$
V	$x^{-m} e^{-r/x}$
VII (I)	$(1 + \frac{x}{a})^{-m}$
IX (I)	$(1 - \frac{x}{a})^m$
XII (I)	$\left[\frac{(3+\beta_1)^{\frac{1}{2}} + \beta_2 + \frac{x}{\sigma}}{(3+\beta_1)^{\frac{1}{2}} + \beta_2 + \frac{x}{\sigma}}\right]^{(\frac{\beta_1}{3+\beta_2})^{\frac{1}{2}}}$
X (III)	e^{-rx}
XI (VI)	x^{-m}

$$\ln y = \int \frac{1}{b_2} \frac{x+a}{(x+b_1/2b_2)^2} dx = \frac{1}{b_2} \ln (x + b_1/2b_2) - \frac{a-b_1/2b_2}{b_2(x+b_1/2b_2)}$$

and

$$y = K(x + b_1/2b_2)^{1/b_2} \exp [\frac{ab_1/2b_2}{b_2(x+b_1/2b_2)}].$$

Thus y is of the type $Kx^{-p}e^{-r/x}$ after a shift of location.

2. Pearson's 1902 Paper

In his 1902 paper, Pearson gave a justification for the use of the method of moments. An outline of his argument is as follows. We seek to fit a curve $y = \varphi(x; c_1,\ldots,c_p)$ to p histogram ordinates, y', where φ is selected to be infinitely differentiable. By Taylor's expansion we can write y as

$$y = \varphi(0) + x\varphi'(0) + \frac{x^2}{2!}\varphi''(0) + \ldots, \tag{3}$$

where the jth derivative of y with respect to x is $\varphi^{(j)}(0) = \varphi^{(j)}(0; c_1,\ldots,c_p)$.

Let $\alpha_j = \varphi^{(j)}(0)$, $j = 0,1,\ldots,p-1$. We can rewrite expression (3) as

$$y = \alpha_0 + \alpha_1 x + \frac{\alpha_2 x^2}{2!} + \ldots + \frac{\alpha_{p-1}x^{p-1}}{(p-1)!} + \varphi^{(p)}(0; c_1,\ldots,c_p)\frac{x^p}{p!} + \ldots, \tag{4}$$

or as

$$y = \alpha_0 + \alpha_1 x + \ldots + \frac{\alpha_{p-1}x^{p-1}}{(p-1)!} + \varphi^{(p)}(0; \alpha_0,\ldots,\alpha_{p-1})\frac{x^p}{p!} + \ldots. \tag{4'}$$

Although he is discussing the fit of y to p histogram ordinates, y', Pearson continues "theoretically" by finding that version of y which minimizes

$$\int (y' - y)^2 dx,$$

which requires $\int (y' - y)\delta y dx = 0$. By use of expression (4'), we can express the change δy in terms of variations in the α's, collect the remaining terms and put

$$R = \frac{x^p}{p!}\varphi^{(p)}(\theta x; \alpha_0,\ldots,\alpha_{p-1}), \qquad |\theta| < 1$$

Hence

$$\int (y' - y)(1 + \frac{\partial R}{\partial \alpha_0})dx\delta\alpha_0 + \ldots + \int (y' - y)(x^{p-1} + \frac{\partial R}{\partial \alpha_{p-1}})dx\delta\alpha_{p-1} = 0. \tag{5}$$

Since the $\delta\alpha$'s are general, expression (5) implies that

$$\int (y' - y)(1 + \frac{\partial R}{\partial \alpha_0})dx = \int (y' - y)(x + \frac{\partial R}{\partial \alpha_1})dx = \ldots = 0.$$

If we let $A\mu_i$ = ith moment of the theoretical curve y and $A'\mu_i'$ = ith moment of the observed curve y', then

$$A\mu_i \simeq A'\mu_i' - \int (y' - y) \frac{\partial R}{\partial \alpha_i} dx. \qquad (6)$$

When the chosen curve is of the correct form and the original sample size for y' is relatively large, we have $(y' - y) \approx 0$ and $\frac{\partial R}{\partial \alpha_i} \approx 0$. Therefore, expression (6) gives rise to $A\mu_i \approx A'\mu_i'$ for $i = 0,\ldots,p$. Now, since the moments are close, the parameters $c_1,\ldots,c_p$ are also close to sample values. This completes Pearson's argument.

Consider a normal distribution $N(c_1,c_2)$. If we let $\mu_1 = c_1$, $\mu_2 = c_2 + c_1^2$, then method of moments would result in the following approximation: $\mu_1 \approx \bar{x}$, $\mu_2 \approx \frac{1}{n}\sum_{i=1}^{n} x_i^2$. Therefore, we only need to solve the system of equations

$$\hat{c}_1 = \bar{x}$$

$$\hat{c}_2 + \hat{c}_1^2 = \frac{1}{n}\sum_{i=1}^{n} x_i^2.$$

This gives

$$\hat{c}_1 = x, \ \hat{c}_2 = s_x^2 = \frac{1}{n}\sum_{i=1}^{n} (x_i - \bar{x})^2.$$

4. Some Comparisons Between MM and ML Estimates

When will MM be approximately the same as ML estimates? In 1922 [CP 18], Fisher explored the possible equivalence of the MM and ML estimates. Suppose we want to fit four parameters with

$$y = \exp[-a^2(x^4 + \rho_1 x^3 + \rho_2 x^2 + \rho_3 x + \rho_4)]. \qquad (7)$$

If the distribution is very close to normal form, then expression (7) takes the form

$$y = c \exp - \frac{x^2}{2\sigma^2} + k_1 \frac{x^3}{\sigma^3} + k_2 \frac{x^4}{\sigma^4} \tag{8}$$

and consequently

$$\frac{\partial}{\partial x} \ln y = \frac{x}{\sigma^2} (1 - 3k_1 \frac{x}{\sigma} - 4k_2 \frac{x^2}{\sigma^2}) \doteq \frac{-x}{\sigma^2 (1 + 3k_1 \frac{x}{\sigma} + 4k_2 \frac{x^2}{\sigma^2})}$$

with k_1, k_2 small. This corresponds to a curve of the Pearson form. Therefore MM estimates are approximately the same as ML estimates when the curve considered is rather close to the normal curve.

The following three examples illustrate calculations of the relative efficiency of MM to ML estimates. In each case $X_1, X_2, \ldots, X_n$ are independently sampled from the given distribution, and we consider the precision of estimates as $n \to \infty$.

Example 1: Uniform. Let X_i have uniform distribution on the interval $[0,\theta]$. Then the MM estimate of θ is $\mu^* = 2\bar{X}$ with variance $\theta^2/3$. The ML estimate, $\hat{\mu}$, is the largest value X_n with variance $n\theta^2/(n+2)(n+1)^2$. Therefore

$$\left\{\left|\frac{\text{Var }(\mu^*)}{\text{Var }(\hat{\mu})}\right|\right\}^{-1} = \left\{\left|\frac{(n+2)(n+1)^2}{3n^2}\right|\right\}^{-1} \to 0 \text{ as } n \to \infty,$$

showing that the MM estimate is relatively useless.

Example 2: Laplace. Let X_i have probability density function (p.d.f.) $\frac{1}{2} \exp(-|x-\theta|)$ with θ unknown. Then the MM estimate of θ is $\mu^* = \bar{X}$ with variance $2/n$, and the ML estimate $\hat{\mu}$ is $X' = \text{median } [X_1, \ldots, X_n]$. Using the asymptotic theory for order statistics, we know that $\text{Var}(X') \approx 1/n$ as n gets large. This result says that the MM estimate is asymptotically twice as variable as the ML estimate.

Example 3: Negative Binomial (Fisher [CP 182]). If we let X_i be the number of failures preceding r successes, then

$$\Pr(X_i = k) = \binom{r+k+1}{k} p^r q^k = \binom{-r}{k} p^r (-q)^k.$$

In discussing the estimation of p and r using a sample of size n for this situation, Fisher actually presents a strong case for MM based on its practical convenience.

The first four central moments of this distribution exist, satisfying one requirement of Pearson's general approach. Then, MM uses the first two moments to solve for p and r, which leads to

$$\hat{p} = \frac{m_2 - \bar{X}}{\bar{X}}, \qquad \hat{r} = \frac{\bar{X}^2}{m_2 - \bar{X}},$$

where m_2 is the sample variance. The variance matrix of $(\hat{p},\hat{r})$ can be obtained from that of $(\bar{X},m_2)$ by the standard delta method, i.e.,

$$\Sigma_{(\hat{p},\hat{r})} = \Sigma_{(\bar{X},m_2)} \left\{ \left| \frac{\partial(\hat{p},\hat{r})}{\partial(\bar{X},m_2)} \right| \right\}^2.$$

With some calculation, it can then be shown that

$$\det \Sigma_{(\hat{p},\hat{r})} = \frac{q^3(r+1)}{N^2},$$

where N is the sample size. Denoting the expected Fisher information matrix by $\mathcal{I}$, we find that

$$\frac{1}{\text{efficiency}} = \frac{\det \Sigma_{(\hat{p},\hat{r})}}{\det \mathcal{I}}$$

$$= 1 + \frac{4}{3}\frac{p}{q(r+2)} + \frac{3p^2}{q^2(r+2)(r+3)} + \cdots$$

Thus, we can conclude that the relative efficiency of MM to ML is less than one and is close to 1 if and only if $\frac{p}{q(r+2)}$ is small.

5. Summary

The fact that the method of moments is easy to use was a major point in its favor before the advent of the computer. Maximum likelihood estimation in practice often led to laborious calculation. Indeed, in his analysis of dilution series data in 1922 [CP 18], Fisher resorted to a modified usage of MM for this reason (see the lecture by Geisser in this volume).

Computer routines now make a maximum likelihood procedure more practical. However, the representation of distributions in Monte Carlo studies when only moments are known borrows from the goals and techniques of Pearson. A further discussion of this development is available in Ord (1972). One modern equivalent of Pearson's problem is that of non-parametric density estimation.

References

Elderton, W.P. (1953). Frequency Curves and Correlation. Washington, D.C.: Harren Press.

Ord, J.K. (1972). Families of Frequency Distributions. London: Griffin.

Pearson, E.S. (1938). Karl Pearson: An Appreciation of Some Aspects of His Life and Work. Cambridge, England: Cambridge University Press.

Pearson, K. (1902). "On the Systematic Fitting of Curves to Observations and Measurements - Part I," Biometrika, 1, 265-303.

FISHER'S CONTRIBUTIONS TO THE ANALYSIS OF CATEGORICAL DATA

Stephen E. Fienberg

1. Introduction

Those who have had only the briefest of introductions to the analysis of categorical data will have learned of Fisher's contributions such as his exact test. But Fisher's work in this area covered a variety of topics, including fundamental papers on the distribution of the chi-square statistic which brought him into a major confrontation with Karl Pearson.

For years Pearson and his students had been applying the chi-square goodness-of-fit test incorrectly to a wide variety of problems. In a series of five papers, beginning in 1922 and continuing through 1928, Fisher wrote about the χ^2 method, correcting Pearson's errors. Specifically he explained how the appropriate degrees of freedom for χ^2 were to be calculated, and why the use of maximum likelihood and other efficient methods of estimation were intimately related to the χ^2 ideas. Of course, these papers were written during the period when Fisher was publishing his landmark papers on statistical estimation and the foundations of statistics [CP 42] (see the related discussion in the presentations by Hinkley in this volume), and thus the link to the more general theory was a natural one.

This presentation will concentrate on the five papers on χ^2, but will also discuss other related papers published by Fisher subsequently. It is interesting to note that the recent literature on loglinear models for categorical data problems can also be traced directly to Fisher, although he did not actually write about methods for multiway tables. In his pioneering paper on $2\times2\times2$ tables, Bartlett (1935) attributes to Fisher the idea of using the equality of Yule's cross-product ratios as a model.

In the next section we begin by reviewing Pearson's work on χ^2.

2. Historical Background

To gain a perspective on Fisher's contributions to the theory of χ^2 and the analysis of categorical data, we need to begin with Pearson's (1900) paper. In that paper, Pearson looked at the problem of comparing a set of observed and expected frequencies through the use of the χ^2 statistic,

$$X^2 = \sum_1^{n'} \frac{(x-m)^2}{m} = \sum_1^{n'} \frac{e^2}{m}, \qquad (1)$$

where m are the expected values (known), x the observed values, n' the number of cells, and e = x-m which are subject to the constraint $\Sigma e = 0$ (i.e., $\Sigma x = \Sigma m$). Pearson assumed that $\underset{\sim}{x}' = (x_1,\ldots,x_{n'})$ has a multinomial distribution which is well-approximated by a multivariate normal. He then showed that X^2 is a quadratic form that has a χ^2 distribution with n' - 1 degrees of freedom. With this result in hand, Pearson went on to consider the case when the m's are not known *a priori*, but are fitted by a model using the data. His argument went roughly as follows.

Let m be the true expected cell value and m_s the corresponding sample-based estimate, such that $m = m_s + \mu$ where μ is considered to be small. Then, if we ignore terms of order $(\mu/m_s)^3$, we have the following approximation to $X^2 - X_s^2$, where $X_s^2 = \sum_1^{n'} (x-m_s)^2/m_s$:

$$X^2 - X_s^2 = -\sum_1^{n'} \frac{\mu(x^2-m_s^2)}{m_s^2} + \sum_1^{n'} \left(\frac{\mu}{m_s}\right)^2 \frac{x^2}{m_s}. \qquad (2)$$

Pearson claimed that this difference is negligible when a large sample is considered. Thus he argued that, in the case of a large sample, X_s^2 also has an approximate χ^2 distribution with n' - 1 degrees of freedom.

This result is obviously wrong. Greenwood and Yule (1915) gave an argument against it by noting that in the approach of comparing two prqportions, the square of the statistic used behaves more like a χ^2 with 1 degree of freedom rather than 3. But it took Fisher to present a more carefully reasoned argument.

3. The 1922 Paper [CP 19]

Fisher in his first paper on χ^2 in 1922 gives the correct solution to the Pearson χ^2 problem with estimated expected frequencies. He begins by noting that the χ^2 test as advocated by Pearson is used in "contingency tables in which the sum of the deviations in any row and column is necessarily zero". He then jumps in and immediately identifies the degrees of freedom in an $r \times c$ table as $(r-1)(c-1)$. Fisher argues that "the values of e can be regarded as independent co-ordinates in generalized space, lying in a subspace of dimension equal to the degrees of freedom due to the linear constraints". The discussion here is not really a "proof" at all, and includes some circular reasoning -- Fisher's intuition, however, was most certainly correct. Fisher also shows that a normal statistic for testing $p_1 = p_2$ in a 2-binomial problem, when squared, is identical to the χ^2 statistic. (Greenwood and Yule conjectured this result, but Fisher was the first to outline a proof.) This result gives some support to his solution of the general χ^2 problem with estimated parameters. Finally, he notes that for a $2 \times s$ table, the associated degrees of freedom is $s-1$. This yields the same result as the correction proposed by Pearson to solve the problem of independence in a $2 \times s$ contingency table. Yet Pearson never really explained why the $2 \times s$ table should be treated differently from the standard $r \times c$ contingency table problem.

In his prefatory note for the Collected Papers, Fisher describes this paper as follows:

> This short paper, with all its juvenile inadequacies, yet did something to break the ice. Any reader who feels exasperated by its tentative and piecemeal character should remember that it had to find its way to publication past critics who, in the first place, could not believe that Pearson's work stood in need of correction, and who, if this had to be admitted, were sure that they themselves had corrected it.

4. The 1923 Paper [CP 31]

While the arguments in the 1922 paper may now seem clear and understandable to us today, they led to a predictable controversy in the 1920s. Thus Fisher elaborated his ideas in [CP 31], stressing the concept of degrees of freedom adjusted for the estimation of parameters. The paper begins with the following cross-classification of problems of interest:

	Random Sample	Selected Sample
True Population	A. No correction needed	B. Correction needed
Reconstructed Population	C. Correction needed	---------

Case A (Pearson's original problem) is the basic situation of multinomial sampling with known parameter values for which there was agreement among all concerned. Case B involves known linear restraints (e.g., fixed row totals in a 2×2 table), and again there was general agreement that an adjustment to the degrees of freedom was needed. Case C involves estimated parameters, and was the point of contention.

Fisher, after outlining the problem, goes on to describe a sampling experiment carried out by Yule (1922) for 2×2 tables, involving 350 observations for Case C. The distribution of the values of X^2 is included here in Table 1.

Table 1: The Distribution of the Values of X^2

	Number Expected, n' = 2	Number Observed	Number Expected, n' = 4
0-0.25	134.02 +	122	10.80 -
0.25-0.50	48.15 -	54	17.58 -
0.50-0.75	32.56 -	41	20.13 -
0.75-1.00	24.21 +	24	21.05 -
1-2	56.00 -	62	80.10 +
2-3	25.91 +	18	63.27 +
3-4	13.22 +	13	45.56 +
4-5	7.05 +	6	31.38 +
5-6	3.86 -	5	21.07 +
6--	5.01 +	5	39.06 +
	349.99	350	350

Fisher points out that "there can be no question that the expectation for n' = 4 completely fails while n' = 2 fits the observations well, and the correction is undoubtedly needed".

4. The 1924 Paper [CP 34]

The argument over the distribution of χ^2 was not settled by Fisher's 1922 and 1923 papers. Pearson (1922) denounced Fisher's claims, without referring specifically to him by name ("I trust my critic will pardon me for comparing him to Don

Quixote tilting at a windmill"). Fisher thus felt compelled to carry on the battle.

In his 1924 paper, Fisher shows with some care what was wrong with Pearson's original reasoning on the distribution of χ^2. He begins by describing three situations where we should not expect to achieve the usual asymptotic χ^2 distribution:

(A) the hypothesis tested is not in fact true,

(B) the method of estimation for the expected values is inconsistent,

(C) the method of estimation employed is inefficient.

Two properties of efficient estimates reviewed by Fisher in this context are worth mentioning here:

(1) The correlation between any two efficient estimates of the same parameter tends to one as the sample size tends to ∞.

(2) The correlation between an efficient and any other consistent estimate is $\sqrt{E}$ where E is the efficiency of the consistent estimate.

In this paper, Fisher also discusses minimizing the value of X^2 with respect to the parameter θ. He notes that the minimum is achieved when

$$\Sigma\left(\frac{x^2-m^2}{m^2}\right)\frac{\partial m}{\partial\theta} = 0. \tag{3}$$

By comparison the maximum likelihood estimate (MLE) satisfies the equations

$$\Sigma\left(\frac{x-m}{m}\frac{\partial m}{\partial\theta}\right) = 0. \tag{4}$$

Fisher claims that, for large samples, the factor $(x+m)/m$, by which the terms in (3) and (4) differ, tends in all cases to the value 2. Hence all methods involving any efficient statistic tend to minimize X^2. He then takes a new statistic, X'^2, equal to $\Sigma(x-m')^2/m'$ where m' is calculated using an efficient estimate, and finds the difference between X^2 and X'^2 as

$$X^2 - X'^2 = \Sigma\left[\frac{(x-m)^2}{m} - \frac{(x-m')^2}{m'}\right]$$

$$= \Sigma[x^2(\frac{1}{m} - \frac{1}{m'})]. \tag{5}$$

Also

$$\frac{1}{m} - \frac{1}{m'} = - \frac{1}{m'^2} \frac{\partial m'}{\partial \theta} \delta\theta + \{\frac{2}{m'^3} (\frac{\partial m'}{\partial \theta})^2 - \frac{1}{m'^2} \frac{\partial^2 m'}{\partial \theta^2}\} (\frac{\delta\theta}{2})^2, \tag{6}$$

where

$$\delta\theta = \theta - \theta' = 0(n^{-\frac{1}{2}}). \tag{7}$$

But since X^2 has been made a minimum, we have

$$\Sigma(\frac{x^2}{m'^2} \frac{\partial m'}{\partial \theta}) = 0.$$

Thus expression (5) reduces to

$$X^2 - X'^2 = (\delta\theta)^2 \Sigma[\frac{1}{m'} (\frac{\partial m'}{\partial \theta})^2]$$

$$= \frac{(\delta\theta)^2}{\sigma^2(\theta')}, \tag{8}$$

and we get a reduction of X^2 when we estimate θ efficiently. Fisher also discussed the effects of estimating θ inefficiently, a research topic which has once again become fashionable in recent years (see the recent discussion in Fienberg, 1979).

5. The 1926 Paper [CP 49]

In his final paper attacking Karl Pearson's use of χ^2, Fisher uses E.S. Pearson's experimental data on the distribution of binomial p, which for each of 12,448 different events contains the frequency of occurrence in two samples of 20 and 15, respectively. Fisher eliminates 780 cases of zero total occurrences, where X^2 is indeterminate, and for the remaining 11,668 cases computes the average X^2 for each value of total number of occurrences. The results are given in Table 2.

Table 2: Average χ^2 for Each Value of Total Number of Occurrences

# of Successes	1	2	3	4	5	6	7	8	9
# of Tables	708	821	779	792	769	730	727	694	630
Total X^2	782.10	834.08	768.82	772.86	807.92	775.74	740.85	697.21	562.33
Average	1.0184	1.0159	0.9869	0.9758	1.0506	1.0497	1.0190	1.0046	0.8926

# of Successes	10	11	12	13	14	15	16	17	Total
# of Tables	643	670	682	668	616	568	524	578	11668
Total X^2	598.09	639.87	707.06	634.06	603.26	618.11	526.52	599.24	11668.12
Average	0.9302	0.9550	1.0368	0.9492	0.9793	1.0882	1.0048	1.0367	1.00001

He notes that in every case the average value is "embarrassingly close" to 1, in no case is it near 3. To this paper, Pearson wrote no reply.

6. The 1928 and 1942 Papers [CP 62, 188]

His dispute with Pearson at least technically behind him, Fisher continued to write about categorical data problems and χ^2, stressing the use of maximum likelihood for fitting the expected values.

In his 1928 paper, Fisher takes a genetic example with underlying cell probabilities $\frac{1}{4}(2+\theta, 1-\theta, 1-\theta, \theta)$ corresponding to observed frequencies (a_1, a_2, a_3, a_4). He notes that $x = a_1 + a_2 - 3(a_3 + a_4)$, $y = a_1 + a_3 - 3(a_2 + a_4)$ each have expectation zero for all values of θ, and that they may be identified with the two degrees of freedom available for testing goodness-of-fit of the model. The appropriate normal quadratic form in x and y, i.e., with their covariance matrix inverse as kernel, is

$$Q^2 = \frac{3}{8n(1-\theta)(1+2\theta)}\{x^2 + y^2 - \tfrac{2}{3}(4\theta-1)xy\}. \tag{9}$$

He then compares Q^2 with the classical chi-square for given θ, namely

$$X^2 = \frac{4}{n}\left\{\frac{a_1^2}{2+\theta} + \frac{a_2^2+a_3^2}{1-\theta} + \frac{a_4^2}{\theta}\right\} - n, \tag{10}$$

the difference being

$$X^2 - Q^2 = \left\{\frac{a_1}{2+\theta} - \frac{a_2+a_3}{1-\theta} + \frac{a_4}{\theta}\right\}^2 \Big/ \frac{(1+2\theta)n}{2\theta(1-\theta)(2+\theta)}.$$

Thus $X^2 - Q^2 > 0$ unless

$$\frac{a_1}{2+\theta} - \frac{a_2+a_3}{1-\theta} + \frac{a_4}{\theta} = 0. \tag{11}$$

But expression (11) is exactly the likelihood equation. Thus the difference $X^2 - Q^2$ can be regarded as that part of the discrepancy between observation and hypothesis which is due to imperfect methods in the estimation of θ.

Fisher goes on to discuss maximum likelihood estimation (MLE) of expected frequencies in general with s cells and r parameters $\theta_1,\theta_2,\ldots,\theta_r$. He notes that the quadratic form analogous to expression (11) is made up of two parts, one of which is a quadratic form distributed in large samples as χ^2_{s-r-1}, and the other being due to errors of measurement, meaning inefficient estimation.

Likelihood estimation for categorical data problems continued to fascinate Fisher, and in 1942 [CP 188] he wrote a brief note on a χ^2 problem for which the likelihood solution comes out neatly. Let a, b, and c be three binomial variates with parameters p, p', and pp', and sample sizes A, B, and C, respectively. Then the value of X^2 with MLE's substituted for the expected values is

$$X^2 = \lambda^2 \left[\frac{A-a}{A(a+\lambda)} + \frac{B-b}{B(b+\lambda)} + \frac{C-c}{C(c-\lambda)}\right], \tag{12}$$

where λ is a root of

$$(A+\lambda)(B+\lambda)(C-\lambda) = (a+\lambda)(b+\lambda)(c-\lambda). \tag{13}$$

Fisher extends this result to the case of s probabilities with s+1 binomial variates (the extra one corresponding to the combined event with probability the product of the s probabilities).

7. Confidence Limits for the Cross-Product Ratio [CP 291]

Even after his retirement to Australia, Fisher continued to write about methods for the 2×2 tables. In one of his last publications [CP 291], he briefly explored the use of the distribution of the exact test statistic (and χ^2 with

Yates' correction) to set limits on the population cross-product ratio. His example was as follows.

Let the observed table be

10	3
2	15

with expected frequencies

10-x	3+x
2+x	15-x

and cross-product ratio

$$\text{cpr} = \frac{(10-x)(15-x)}{(3+x)(2+x)} .$$

Now

$$X_c^2 = (x-\tfrac{1}{2})^2 \left(\frac{1}{10-x} + \frac{1}{3-x} + \frac{1}{2-x} + \frac{1}{15-x}\right).$$

Next we pick x to make $X^2 = 3.841$ (the 95th percentile), yielding $x = 3.0491$ and cpr = 2.720. The latter, Fisher argues, is an "upper limit" for the true cpr.

8. The Exact Test and χ^2

Fisher introduced his exact test for 2×2 tables with the now classic example of the lady tasting tea in The Design of Experiments [DOE, 1935] (see the discussion of this example in the lecture by Holschuh in this volume). He advocated the use of the exact test in subsequent issues of Statistical Methods for Research Workers [SMRW, 1925] (Section 21.02), referring to the use of χ^2 as an approximation. The χ^2 test for 2×2 tables with the correction for continuity introduced by Yates (used in Section 7 above) was an attempt to get tail probability values that conformed more closely to those of the exact test than did those from the uncorrected χ^2 statistic.

In a 1941 Science article [CP 183], in response to a paper by E.B. Wilson, Fisher tried to clarify why he believed that the exact test should be used in 2-binomial experiments when the sample sizes are small. The discussion in this paper is obscure at best, and to the present day few have correctly described Fisher's position in a coherent fashion. Indeed his position on this issue (as on others) seems to have changed over time. Berkson (1978) and Kempthorne (1979) have continued the debate over the appropriateness of the exact test, and I fear that we will continue to see papers on this topic in the future.

My current judgment is that Fisher and others consistently overstated the dangers of using X^2 in small samples as if it really was distributed as a χ^2 variate (e.g., see the small sample studies of Larntz, 1978).

References

Bartlett, M.S. (1935). "Contingency Table Interactions," Supplement to the Journal of the Royal Statistical Society, 2, 248-252.

Berkson, J. (1978). "In Dispraise of the Exact Test," Journal of Statistical Planning and Inference, 2, 27-42.

Fienberg, S.E. (1979). "The Use of Chi-Squared Statistics for Categorical Data Problems," Journal of the Royal Statistical Society, Series B, 41, 54-64.

Greenwood, M. and G.U. Yule (1915). "The Statistics of Antityphoid and Anti-cholera Inoculations, and the Interpretation of Such Statistics in General," Proceedings of the Royal Society of Medicine, Section of Epidemiology and State Medicine, viii, 113.

Kempthorne, O. (1979). "In Dispraise of the Exact Test: Reactions," Journal of Statistical Planning and Inference, 3, 199-213.

Larntz, K. (1978). "Small-Sample Comparisons of Exact Levels for Chi-Squared Goodness-of-Fit Statistics," Journal of the American Statistical Association, 73, 253-263.

Pearson, K. (1900). "On the Criterion that a Given System of Deviations from the Probable in the Case of a Correlated System of Variables is Such that it Can be Reasonably Supposed to have Arisen from Random Sampling," The London, Edinburgh, and Dublin Philosophical Magazine and Journal of Science, 1, 157.

Pearson, K. (1922). "Further Note on the χ^2 Test of Goodness of Fit," Biometrika, 14, 418.

Yule, G.U. (1922). "On the Application of the χ^2 Method to Association and Contingency Tables with Experimental Illustration," Journal of the Royal Statistical Society, 5, 95-104.

This research was supported by Office of Naval Research Contract N00014-78-C-0600 to the University of Minnesota.

THEORY OF STATISTICAL ESTIMATION: THE 1925 PAPER

David Hinkley

1. Introduction

As Fisher himself suggests, the 1925 paper [CP 42] is a compact refinement of the 1922 paper [CP 18]. It is short, sometimes terse, but monumental in concept. It was, I believe, largely ignored for more than thirty years.

Criticism of the 1922 paper (not just by Pearson apparently) occurred because of Fisher's notion of statistics having sampling distributions in a series of hypothetical repeated experiments. The 1925 paper has a prologue with an unclear account of what Fisher means by "infinite hypothetical population". These ideas will be discussed in a later lecture, but let me note that in 1935 Fisher said, "my own definition of probability is not based on the limit of experimental frequencies" [CP 124].

2. Comments on the 1922 Paper

Fisher said later, in a discussion of the 1935 Journal of the Royal Statistical Society paper [CP 124],

> I ought to mention that the theorem that if a sufficient statistic exists, then it is given by the method of maximum likelihood was proved in my paper of 1921 [1922]. ... It was this that led me to attach especial importance to this method. I did not at that time, however, appreciate the cases in which there is no sufficient statistic, or realize that other properties of the likelihood function, in addition to the position of the maximum, could supply what was lacking. [CP 124, p. 82].

Actually, the 1922 paper is somewhat confusing on the key notions of sufficiency and efficiency. For example, intrinsic accuracy (essentially expected information) is defined [CP 18, p. 310] as "the weight [inverse variance] in large samples, divided by the number in the sample, of that statistic of location which satisfies the criterion of sufficiency"; later, in the version of this paper reprinted in the Collected Papers, Fisher changed the last word to "efficiency". There is an explicit assumption in 1922 that maximum likelihood is optimal, and some passages suggest it to be sufficient. Using normal approximations, Fisher shows that if T_1 is a sufficient estimate, then any other estimate cannot have smaller variance. Sufficiency is defined as $f_\theta(T_1,T_2) = f_\theta(T_1)g(T_2|T_1)$ for any T_2, but the now-familiar factorization theorem is missed because of the emphasis on estimation. Then, later, "we require a method which will automatically lead us to the statistic by which the criterion of sufficiency is satisfied. Such a method is, I believe, provided by the Method of Maximum Likelihood, although I am not satisfied ...". The "proof" that maximum likelihood is sufficient is, of course, true in the converse sense, and interestingly Fisher says "any statistic which fulfills the condition of sufficiency must be a solution obtained by the method of the optimum [i.e., maximum likelihood (ML)]". One consequence of believing ML to be sufficient is that the smallest variance possible is claimed to be one over the expected information, because of the earlier efficiency result. Most of this confusion disappears by 1925.

3. Outline of the 1925 Paper

After the prologue, the 1925 paper opens with comments about consistency (both weak and Fisher versions), and efficiency. Fisher was greatly interested in efficiency, and this has, I believe, resulted in misplaced stress on Fisher information in connection with standard errors for MLE by later writers. Another important point for the modern reader to note is that Fisher was often happy to take large-sample distributions as exactly normal in theoretical calculations -- but only where this was correct for the order of the required result. This comes up in the "proof" that if T is efficient, then corr(T,T') $= \sqrt{E}$ where E = efficiency of T' Fisher refers to $T-\theta$ as sampling error, and T'-T as estimation error, which he relates to chi-square goodness-of-fit tests.

The fifth section of the 1925 paper deals with what we now call the method of scoring, which Pearson derided, where the likelihood equation is solved by linear expansion and iteration. Detailed comments about this appear in the lecture by Runger.

The remainder of the 1925 paper deals largely with information loss, and it is here that some grand ideas emerge, so let us look carefully at these parts of the paper.

4. Information Loss

Information loss was an important consideration because Fisher now realized that asymptotic theory was not reliable for finite sample sizes. Certainly he was very impressed by "Student's" 1908 exact result on the t statistic. Other oddities also were tied to the failure of asymptotics. For example, the efficient minimum chi-square estimation method had not worked on grouped normal data, as Fisher showed in the 1922 paper [CP 18]. So, the small sample question was: "How much information did ML lose, and how can it be recovered?" The next question would be: "How do I use the recovered information?"

To use modern notation, suppose that for a single parameter θ, $\ell_\theta(x)$ is the log likelihood, $\ell_\theta(x) = \log f_\theta(\text{observation})$, whose derivatives with respect to θ are $\dot{\ell}_\theta$ and $\ddot{\ell}_\theta$. The intrinsic accuracy, or amount of information, is

$$\mathcal{I}_\theta = E(\dot{\ell}_\theta^2) = E(-\ddot{\ell}_\theta),$$

from which the additive property follows.

Some of Fisher's insight comes in the demonstration of the efficiency of maximum likelihood. Here we denote total sample log likelihood by ℓ_θ^S, say, and $\hat{\theta}$ satisfies $\dot{\ell}_{\hat{\theta}}^S = 0$. It is then assumed that $-\ddot{\ell}_{\hat{\theta}}^S \to \mathcal{I}_\theta^S$, so that Taylor expansion of $\dot{\ell}_\theta^S$ gives

$$\dot{\ell}_\theta^S = -\mathcal{I}_\theta^S(\hat{\theta}-\theta) \qquad \text{for small } |\theta-\hat{\theta}|. \tag{1}$$

Now suppose that T is another statistic, and let the log likelihood based only on T be, in Fisher's discrete notation,

$$\ell^T = \sum_{\{s:T=t\}} \ell_\theta^S.$$

Further, if T is asymptotically normal with variance σ^2 then[1]

$$\frac{1}{\sigma^2} = -\ddot{\ell}_\theta^T.$$

---

[1]Fisher's notation does not distinguish between evaluations at θ and $\hat{\theta}$, and the latter may sometimes be intended.

"The problem of making σ^2 as small as possible, is the problem of so grouping the several sorts of samples under the same values ... of T, that the second differential coefficient be as large as possible." Then

$$\frac{1}{\sigma^2} = -\ddot{\ell}_\theta^T = \frac{\Sigma\, \dot{\ell}_\theta^S f_\theta^S}{\Sigma\, f_\theta^S} - \frac{\Sigma \frac{d}{d\theta}(\dot{\ell}_\theta^S f_\theta^S)}{\Sigma\, f_\theta^S},$$

which by the linear approximation (1) comes down to

$$[E\{\dot{\ell}_\theta^S|T\}]^2 - E\{\ddot{\ell}_\theta^S|T\} - E\{(\dot{\ell}_\theta^S)^2|T\}$$

$$= (\mathcal{I}_\theta^S)^2[E\{(\hat{\theta}-\theta)|T\}]^2 + \mathcal{I}_\theta^S + (\mathcal{I}_\theta^S)^2 E\{(\hat{\theta}-\theta)^2|T\}.$$

The first and third terms are non-negative, so that $\hat{\theta} = T$ gives the minimum. Loss of efficiency is thus traced to variation of $\hat{\theta}$ over the set fixed by T. The question remains, I think, in Fisher's mind: does this apply when T is also efficient? The answer is affirmative, as we shall see.

There is next a section on weighting, to which I shall return later. Fisher then focusses on small-sample problems and discusses sufficiency. Here let T_1 be sufficient and let T_2 be another statistic. Then

$$f_\theta(T_1,T_2) = g_\theta(T_1)h(T_1,T_2).$$

"If this is fulfilled for all T_2, then T_1 is sufficient". Further, it is a consequence that, for some φ,

$$T_1(x_1,\ldots,x_n) = \varphi\{T_1(x_1,\ldots,x_p),T_1(x_{p+1},\ldots,x_n)\}.$$

Recognition of exponential families as a solution to this requirement is several years away.

The next step is to consider $\{x : T = t\}$ and to see how much information is lost in the reduction to t. If g_θ^T is the frequency function of T, then

$$\mathcal{I}_\theta^T = \sum_t \frac{(\dot{g}_\theta^T)^2}{g_\theta^T} \quad \text{as opposed to} \quad \mathcal{I}_\theta^S = \sum_x \frac{(\dot{f}_\theta^S)^2}{f_\theta^S}.$$

The effect of amalgamation is, therefore, to reduce the information by

$$\sum_t \sum_{x|t} \left\{ \left(\frac{\dot{f}^S_\theta}{f^S_\theta} \right)^2 f^S_\theta - \left(\frac{\dot{g}^T_\theta}{g^T_\theta} \right)^2 g^T_\theta \right\} = \sum_x f^S_\theta \left\{ \frac{\dot{f}^S_\theta}{f^S_\theta} - \frac{\dot{g}^T_\theta}{g^T_\theta} \right\}^2 , \tag{2}$$

which is non-negative, and zero only if $\dot{\ell}^S_\theta$ = constant on $\{x : T = t\}$. Hence $\{x : \dot{\ell}^S_\theta = \text{constant}\}$ defines a mathematically sufficient partition, and if the set is independent of θ, then the sufficient partition gives an observable sufficient statistic. In particular, $\hat{\theta} = \dot{\ell}^S_{\hat{\theta}} = 0$ will be sufficient only if $\{x : \dot{\ell}^S_\theta = \text{constant}\}$ is independent of θ. Otherwise, expression (2) is strictly positive for any statistic (i.e., a function of x alone).

Thus sufficiency is tied to $\dot{\ell}^S_\theta$ and Fisher information. This leads to the theory of information loss in small samples. There has already been a virtuoso display of computational mathematics in finding $\mathcal{I}^{\text{median}}_\theta$ for the Cauchy and Laplace distributions, where Fisher notes the slow convergence to the limit. For the Laplace distribution "the example stresses the importance of investigating the actual behavior of statistics in finite samples, instead of relying wholly upon their calculated behaviour in infinitely large samples".

Fisher's next calculation revolves around the multinomial, perhaps because of his interest in χ^2 and genetics problems. We can be slightly more general. I shall continue to follow Fisher's practice of ignoring "nonessential" remainder terms. First, the Taylor expansion of the efficient score gives

$$\dot{\ell}^S_\theta = \dot{\ell}^S_{\hat{\theta}} + (\theta-\hat{\theta})\ \ddot{\ell}^S_{\hat{\theta}}. \tag{3}$$

We have just traced the information loss to variation of $\dot{\ell}^S_\theta$ over $\{x : \dot{\ell}^S_\theta = 0\}$, which by expression (3) may be expressed by the equations

$$\mathrm{Var}(\dot{\ell}^S_\theta|\hat{\theta}) = (\theta-\hat{\theta})^2\ \mathrm{Var}(\ddot{\ell}^S_{\hat{\theta}}|\hat{\theta}) \tag{4}$$

and

$$\mathcal{I}^S_\theta - \mathcal{I}^{\hat{\theta}}_\theta = E_{\hat{\theta}}\ \mathrm{Var}(\dot{\ell}^S_\theta|\hat{\theta}). \tag{5}$$

We can see this latter identity by noting that $\dot{\ell}_\theta^{\hat{\theta}} + \dot{\ell}_\theta^{S|\hat{\theta}} = \dot{\ell}_\theta^S$, and then computing

$$\mathcal{I}_\theta^S = E_{S|\hat{\theta}} E_{\hat{\theta}} \{\dot{\ell}_\theta^{\hat{\theta}} + \dot{\ell}_\theta^{S|\hat{\theta}}\}^2.$$

There are some slippery steps now, which Fisher negotiates carefully. I drop the superscript S. The final term in expression (4) is

$$\begin{aligned} \operatorname{Var}(\ddot{\ell}_{\hat{\theta}}|\hat{\theta}) &= \operatorname{Var}\{\ddot{\ell}_{\hat{\theta}}|S : \dot{\ell}_{\hat{\theta}} = 0\} \\ &= \operatorname{Var}\{\ddot{\ell}_\theta|S : \dot{\ell}_\theta = 0\} \\ &= \operatorname{Var}(\ddot{\ell}_\theta) - \frac{\{\operatorname{Cov}(\dot{\ell}_\theta,\ddot{\ell}_\theta)\}^2}{\operatorname{Var}(\dot{\ell}_\theta)} . \end{aligned} \tag{6}$$

Together with expressions (4) and (5), this last result shows that, ignoring terms that tend to zero,

$$\mathcal{I}_\theta^S - \mathcal{I}_\theta^{\hat{\theta}} = \mathcal{I}_\theta^S(\gamma_\theta^S)^2,$$

where $(\gamma_\theta^S)^2$ is the invariant subsequently defined by Efron (1975) as

$$[\operatorname{Var}\ \ddot{\ell}_\theta - \{\operatorname{Cov}(\dot{\ell}_\theta,\ddot{\ell}_\theta)\}^2/\operatorname{Var}\ \dot{\ell}_\theta]/(\mathcal{I}_\theta^S)^2.$$

A similar argument, not given by Fisher, leads to the statement

$$\mathcal{I}_\theta^T = \mathcal{I}_\theta^{\hat{\theta}} - (\mathcal{I}_\theta^S)^2 \operatorname{Var}(T-\hat{\theta}).$$

Fisher's demonstration would probably look like this:

$$\dot{\ell}_\theta^S = \ell_{\hat{\theta}}^S + (\theta - T + T - \hat{\theta})\ddot{\ell}_{\hat{\theta}}^S,$$

so that using expression (3) we have

$$\begin{aligned} \operatorname{Var}(\dot{\ell}_\theta^S|T) &= E_{\hat{\theta}|T} \operatorname{Var}(\dot{\ell}_\theta^S|T,\hat{\theta}) + \operatorname{Var}_{\hat{\theta}|T} E(\dot{\ell}_\theta^S|T,\hat{\theta}) \\ &= E\{(\hat{\theta}-\theta)^2|T\} \operatorname{Var}(\ddot{\ell}_{\hat{\theta}}^S|T,\hat{\theta}) + \operatorname{Var}_{\hat{\theta}|T}\{(\theta - T + T - \hat{\theta})\ddot{\ell}_{\hat{\theta}}^S|T\}, \end{aligned}$$

and hence

$$E \operatorname{Var}(\dot{\ell}_\theta^S|T) = \mathcal{I}_\theta^S - \mathcal{I}_\theta^{\hat{\theta}} + (\mathcal{I}_\theta^S)^2 \operatorname{Var}(T-\hat{\theta}).$$

What does this have to do with precision of estimation? Fisher does not seem very concerned with this question; rather I think he was concerned with showing that: (a) $\mathcal{I}_\theta^{\hat{\theta}} > \mathcal{I}_\theta^T$ for any efficient T, and (b) the discrepancy $\mathcal{I}_\theta^S - \mathcal{I}_\theta^{\hat{\theta}}$ is due to having lost $-\ddot{\ell}_{\hat{\theta}}^S$ (look back at the calculation). He thought of $-\ddot{\ell}_{\hat{\theta}}^S = I$ as being the true weight to attach to $\hat{\theta}$, and goes on to establish this directly as follows.

Suppose that we have $\hat{\theta}_1,\ldots,\hat{\theta}_m$ derived from samples $S_1,\ldots,S_m$ of equal size n (large), and that $\hat{\theta}$ denotes the overall MLE from the combined sample S, the union of $S_1,\ldots,S_m$. Then

$$0 = \dot{\ell}_{\hat{\theta}}^S = \sum_{j=1}^m \dot{\ell}_{\hat{\theta}}^{S_j} = \sum_{j=1}^m \ddot{\ell}_{\hat{\theta}_j}^{S_j}(\hat{\theta} - \hat{\theta}_j),$$

implying $\hat{\theta} = \Sigma\hat{\theta}_j I_j/\Sigma I_j$, where $I_j = -\ddot{\ell}_{\hat{\theta}_j}^{S_j}$. In an earlier calculation, Fisher has shown that if $\mathrm{Var}(\hat{\theta}_j) = w_j^{-1}$, where $w_j = E(w_j) + e_j = w_j' + e_j$ with the e_j being random errors of weighting, then the difference in precisions of the weighted averages

$$T(w) = \Sigma w_j\hat{\theta}_j/\Sigma w_j \quad \text{and} \quad T(w') = \Sigma w_j'\hat{\theta}_j/\Sigma w_j'$$

is expressible as

$$\frac{1}{\mathrm{Var}\{T(w)\}} - \frac{1}{\mathrm{Var}\{T(w')\}} = \sum_{j=1}^m \frac{\mathrm{Var}(w_j)}{w_j'}$$

for large m. Now, if we let $w_j = I_j$ and $w_j' = \mathcal{I}_\theta^{S_j}$, then for $\bar{\hat{\theta}} = m^{-1}\Sigma\hat{\theta}_j$,

$$\frac{1}{\mathrm{Var}(\hat{\theta})} - \frac{1}{\mathrm{Var}(\bar{\hat{\theta}})} = \sum_{j=1}^m \frac{\mathrm{Var}(I_j)}{\mathcal{I}_\theta^{S_j}},$$

i.e.,

$$\frac{1}{\mathrm{Var}(\hat{\theta})} - \frac{m}{\mathrm{Var}(\hat{\theta}_1)} = m\,\mathcal{I}_\theta^{S_1}(\gamma_\theta^{S_1})^2, \qquad (7)$$

the last step following from our earlier calculation of $\mathrm{Var}(\ddot{\ell}_{\hat{\theta}}^S)$ in expression (6). Essentially Fisher has shown here that:

(i) $\hat{\theta}$ for all data is a weighted average of $\hat{\theta}_j$ with weights I_j,

(ii) weights $\mathcal{J}_\theta^{S_j}$ are <u>not</u> equivalent to weights I_j,

(iii) the loss of precision using $\mathcal{J}$ in place of I corresponds exactly to the loss of information in reducing S to $\hat{\theta}$.

This points to (a) the joint sufficiency of $(\hat{\theta}, I)$, and (b) the use of $I^{-\frac{1}{2}}$ as a standard error for $\hat{\theta}$.

We might actually try to use expression (7) to compute $\mathrm{Var}(\hat{\theta}_1)$, say. Suppose that

$$\mathrm{Var}(\hat{\theta}_1) = \frac{1}{ni_\theta}\left(1 + \frac{\delta_\theta}{n}\right), \tag{8}$$

where

$$i_\theta = n^{-1} \mathcal{J}_\theta^{S_1} = \text{unit information.}$$

Then expression (7) says that

$$mni_\theta\left(1 - \frac{\delta_\theta}{mn}\right) - mni_\theta\left(1 - \frac{\delta_\theta}{n}\right) = mni_\theta (\gamma_\theta^{S_1})^2 ,$$

which is valid for large m. Letting $m \to \infty$ we have

$$\delta_\theta = n(\gamma_\theta^{S_1})^2 . \tag{9}$$

This is actually correct, i.e., provides $\mathrm{Var}(\hat{\theta}_1)$ to second order, as long as $E\hat{\theta}_1 = \theta$. Thus expression (9) is correct in the location case. (To see that the proof is wrong in general, suppose that T is the natural sufficient statistic for an exponential family and let $\theta = E(T)$, so that $\hat{\theta} = T$. Then $\mathrm{Var}(\hat{\theta}) = cn^{-1}$ exactly. Now if $\varphi = g\{E(T)\}$, then $\mathrm{Var}(\varphi) = c'n^{-1} + dn^{-2} \ldots$ with $d \neq 0$. But γ_θ is invariant under reparameterization, so that if expression (7) is correct d would have to be zero.)

Information calculations do not correspond to variance calculations, i.e., $\mathrm{Var}(T) = 1/\mathcal{J}_\theta^T$ is not correct except in the location case. The discrepancy, however, is independent of the method of estimation, as Efron (1975) shows. This implies a stronger version of Fisher's result for efficient T, namely that

$$\text{Var}(T) = \text{Var}(\hat{\theta}) + \text{Var}(T-\hat{\theta}) + O(n^{-2}),$$

or

$$\text{corr}(\hat{\theta}, T-\hat{\theta}) = O(n^{-1}).$$

The developments that we have just described have been followed up in the impressive work of C.R. Rao (1961, 1962, 1963) and Efron (1975). (See also Efron and Hinkley, 1978.) Efron's 1975 paper is based on Fisher's geometric ideas, and gives these ideas fuller expression in terms of curved exponential families. The importance of Fisher's second-order calculations are confirmed in the example of maximum likelihood estimation of a Cauchy location parameter, for which some numerical results are cited in the next lecture.

5. Further Comments

It is interesting that Fisher rarely if ever discussed large-sample tests of hypotheses about θ. Most of us are familiar with the test statistics

$$\sqrt{V} = \dot{\ell}^S_{\theta_0} / \sqrt{\mathcal{I}^S_{\theta_0}}, \qquad W = 2(\ell^S_{\hat{\theta}} - \ell^S_{\theta_0}),$$

both V and W being approximately χ^2_1 in large samples when $\theta = \theta_0$. Neyman, Pearson, and Rao are associated with V; Wilks is associated with W. It is likely that Fisher did not consider such tests seriously because he would have constructed parametric tests from fiducial distributions of parameters; very roughly speaking he would have computed $\text{pr}(\theta > \theta_0 | S)$. His large-sample theory was used for finding an approximate sufficient statistic, from which a fiducial distribution could be derived. Both V and W spring out of the Neyman-Pearson testing theory, where alternative hypotheses are used as mathematical devices for arriving at unique "inferences" about experimental hypotheses. Fisher did not seem to think much of this approach. Actually $\sqrt{V}$ is not an efficient statistic in Fisher's sense (see Efron, 1975). We shall return to $\sqrt{V}$ and W in the lecture on conditional inference. Note that W is a reasonable "Fisherian" statistic in the sense that it is based only on the likelihood values.

In the next lecture Runger looks at the method of scoring and shows some numerical results relating to Fisher's information-loss theory.

References

Efron, B. (1975). "Defining the Curvature of a Statistical Problem (With Applications to Second Order Efficiency) (With Discussion)," Annals of Statistics, 3. 1189-1242.

Efron, B. and D.V. Hinkley (1978). "Assessing the Accuracy of the Maximum Likelihood Estimator: Observed Versus Expected Fisher Information," Biometrika, 65, 457-487.

Rao, C.R. (1961). "Asymptotic Efficiency and Limiting Information." In Proceedings of Fourth Berkeley Symposium on Mathematical Statistics and Probability (Jerzy Newman, ed.). Berkeley: University of California Press, 532-545.

Rao, C.R. (1962). "Efficient Estimates and Optimum Inference Procedures in Large Samples," Journal of the Royal Statistical Society, Series B, 24, 46-72.

Rao, C.R. (1963). "Criteria of Estimation in Large Samples," Sankhyā, 25, 189-206.

This research was supported by National Science Foundation Grant MCS 7904558 to the University of Minnesota.

SOME NUMERICAL ILLUSTRATIONS OF FISHER'S THEORY OF STATISTICAL ESTIMATION

George Runger

1. Introduction

In the 1925 paper [CP 42] discussed in the previous lecture, Fisher redefines the efficiency of a statistic, T, as the limiting value of the ratio $\mathcal{I}^T/\mathcal{I}^S$, where $\mathcal{I}^T$ is the information contained in T while $\mathcal{I}^S$ is that in the sample. He discusses efficiency both in small and large samples, and shows that if no sufficient statistic exists, then some loss of information will necessarily ensue upon the substitution of a single estimate for the original data upon which it was based.

Fisher demonstrates that maximum likelihood (ML) will produce an efficient statistic if one exists. "But since the equations of ML do not always lend themselves to direct solution, it is of importance that starting with an inefficient estimate, we can by a single process of approximation obtain an efficient one." For this purpose, Fisher introduces the method of scoring as a means of obtaining an efficient estimate for the Cauchy location parameter. This is the first topic of the present lecture. We will also illustrate Fisher's second-order results for variance of the maximum likelihood estimate (MLE).

2. Methods of Scoring

Suppose $X_1,\ldots,X_n$ is a random sample with common Cauchy probability density function (p.d.f.) $f_\theta(x) = \dfrac{1}{\pi[1+(x-\theta)]^2}$, and that T_1 is an estimate of θ based on $X_1,\ldots,X_n$. Then, with notation defined as in the preceding lecture, we have

$$\dot{\ell}^S_\theta = 2\,\Sigma\ \frac{(x-\theta)}{1+(x-\theta)^2},$$

$$\ddot{\ell}^S_\theta = 2\,\Sigma\ \frac{(x-\theta)^2-1}{[1+(x-\theta)^2]^2},$$

and

$$E(-\ddot{\ell}_\theta^S) = \mathcal{I}_\theta^S = \frac{n}{2}.$$

Expand $\dot{\ell}_\theta$ about $\theta = T_1$ as $\dot{\ell}_\theta = \dot{\ell}_{T_1} + (\theta - T_1)\ddot{\ell}_{T_1}$ (ignoring the remainder terms) and substitute $\theta = \hat{\theta}$, the MLE for θ. With $\dot{\ell}_{\hat{\theta}} = 0$ we can solve the resulting equation to get

$$\hat{\theta} = T_1 + \frac{\dot{\ell}_{T_1}^S}{-\ddot{\ell}_{T_1}^S}. \tag{1}$$

The idea is then to take $T_2' = T_1 + \dot{\ell}_{T_1}/-\ddot{\ell}_{T_1}$ as a second-step estimate; this corresponds to the Newton-Raphson method of solving $\dot{\ell}_\theta = 0$. Fisher, however, recommends the use of $T_2 = T_1 + \dot{\ell}_{T_1}^S/\mathcal{I}_\theta^S$ for its computational simplicity.

One can proceed with T_1 replaced by T_2 in expression (1) and obtain T_3, etc. If T_1 is suitably chosen so that $\mathrm{Var}(T_1 - \hat{\theta}) = cn^{-1}$, then $\mathrm{Var}(T_2 - \hat{\theta}) = dn^{-2}$. Fisher concludes, in his paper, that the differences in the information contents of $\hat{\theta}$ and the estimates obtained by scoring are $\mathcal{I}^{\hat{\theta}} - \mathcal{I}^{T_2} = o(1)$ and $\mathcal{I}^{\hat{\theta}} - \mathcal{I}^{T_3} = o(n^{-1})$. Presumably the same equations will be true for T_2' and T_3'.

Some numerical results on the efficiency of scoring in ML estimation of Cauchy location with $n = 15$ are shown in Table 1, where empirical variances from 5000 Monte Carlo samples are given (see the Appendix for details).

3. Variance of the MLE

Using Fisher's theory with a modification by Efron (1975) -- see also Haldane and Smith (1956) and Rao (1963) -- we have that

$$\mathrm{Var}(\hat{\theta}) = \frac{1}{\mathcal{I}^{\hat{\theta}}} = \frac{1}{\mathcal{I}^S}\left(1 + \frac{\gamma^2}{n}\right), \tag{2}$$

ignoring terms of order n^{-2}. For the Cauchy distribution $\mathcal{I}^S = \frac{n}{2}$ and $\gamma^2 = \frac{5}{2}$. Table 2 compares the second-order result (2) with Monte Carlo estimates of $\mathrm{Var}(\hat{\theta})$, showing better approximation than is achieved by the asymptotic result $1/\mathcal{I}_\theta^S$.

Table 1: Numerical Results for the Method of Scoring in Obtaining MLE of Cauchy Location

Estimate*	Empirical Variance (Monte Carlo standard error)	Theoretical Variance (Source)
T_1	0.1973 (.0045)	0.1838 (Fisher)
T_2'	0.1666 (.0037)	
T_2	0.1674 (.0038)	
T_3	0.1643 (.0036)	
T''	0.1617 (.0034)	0.1556 (Fisher-Efron)
$T'' - T_1$	0.041 (.0030)	0.0282
$T'' - T_2$	0.0095 (.0018)	
$T'' - T_3$	0.0048 (.0015)	0

*Notation: $\dot{\ell}_T$ = first derivative of log likelihood at $\theta = T$.

$\ddot{\ell}_T$ = second derivative of log likelihood at $\theta = T$.

$\mathcal{J}_T = E(-\ddot{\ell})$ = expected Fisher information = $\frac{1}{2}$ n in the Cauchy case.

Estimates: T_1 = median, $T_2 = T_1 + \dot{\ell}_{T_1}/\mathcal{J}_{T_1}$.

$T_2' = T_1 + \dot{\ell}_{T_1}/(-\ddot{\ell}_{T_1})$, $T_3 = T_2 + \dot{\ell}_{T_2}/\mathcal{J}_{T_2}$.

$T'' = \hat{\theta}$ as computed by iterative re-weighted least squares.

Table 2: Variance of MLE for Cauchy Location

Sample size n	9	10	11	13	15	19	20
Asymptotic variance formula, $1 \div \mathcal{J}^S$	0.222	0.200	0.182	0.154	0.133	0.105	0.100
Fisher-Efron theory, expression (2)	0.282	0.250	0.222	0.184	0.156	0.119	0.1125
Monte Carlo (standard error)	0.332 (.02)	0.270 (.105)	0.247 (.02)	0.189 (.015)	0.158 (.01)	0.117 (.01)	0.113 (.005)

Source: Efron (1975).

4. The Use of Observed Information as Weight

The last part of Fisher's 1925 paper [CP 42] deals with the precision to be attached to $\hat{\theta}$. Suppose there are m samples of size n. Let the log likelihood function for the jth sample be $\ell_{j,\theta}$ with MLE based on $\ell_{j,\theta}$ denoted by $\hat{\theta}_j$. If the overall pooled log likelihood is ℓ_θ with MLE $\hat{\theta}$, then we have

$$0 = \dot{\ell}_{\hat{\theta}} = \sum_1^m \dot{\ell}_{j,\hat{\theta}}$$

$$\doteq \sum_1^m \dot{\ell}_{j,\hat{\theta}_j} + \sum_1^m (\hat{\theta}-\hat{\theta}_j)\ddot{\ell}_{j,\hat{\theta}_j}$$

$$= \sum_1^m (\hat{\theta}-\hat{\theta}_j)\ddot{\ell}_{j,\hat{\theta}_j} .$$

Solving for $\hat{\theta}$ gives the equation

$$T_1 = \frac{\sum_1^m \hat{\theta}_j I_j}{\sum_1^m I_j},$$

where

$$I_j = -\ddot{\ell}_{j,\hat{\theta}_j}$$

is the observed information in the jth sample.

Another way to obtain an estimate of θ is simply to take the arithmetic average of the m $\hat{\theta}_j$'s, i.e., $T_2 = \frac{1}{m} \Sigma\, \hat{\theta}_j$. Fisher's theory, as amplified by Efron (1975), suggests that for large samples

$$\mathrm{Var}(T_2) = \frac{1}{m}\mathrm{Var}(\hat{\theta}_1) = \frac{1}{mni_\theta}\left(1 + \frac{\gamma^2}{n}\right),$$

while

$$\mathrm{Var}(T_1) \approx \mathrm{Var}(\hat{\theta}) = \frac{1}{mni_\theta}\left(1 + \frac{\gamma^2}{mn}\right),$$

where i_θ is the information contained in one observation. Table 3 below gives some numerical results on the efficiency of pooled MLEs for estimating Cauchy location. Empirical variances are from 1250 sets of four samples of size n = 15. The numerical results confirm the superiority of weighting with observed information.

Table 3: Comparison of Two Methods of Weighting MLEs

Estimate*	Empirical Variance (standard error)	Theoretical Variance**
T_2	0.0392 (0.0017)	0.0389
T_1	0.0348 (0.0015)	
$\hat{\theta}$	0.0339 (0.0014)	0.0347
$\hat{\theta} - T_2$	0.00489 (0.0006)	0.0042
$\hat{\theta} - T_1$	0.00115 (0.0002)	0

*For estimates $\hat{\theta}_j$ $(j = 1,\ldots,4)$ and $I_j = -\ddot{\ell}_{\hat{\theta}}$ in the jth sample, $T_2 = \frac{1}{4}\Sigma\,\hat{\theta}_j$, $T_1 = \Sigma\, I_j\hat{\theta}_j/\Sigma\, I_j$ and $\hat{\theta}$ is obtained by iterative re-weighted least squares on the pooled sample.

**Ignoring terms of order n^{-2}.

Appendix

All Monte Carlo computations for this lecture were carried out on a CDC 6400 at the University of Minnesota. Each Cauchy variable was generated as a ratio of standard normals, normal deviates being obtained from subprogram NORMAL contained in the 6400's FORTRAN subprogram library. This subprogram is based on the composition method (Marsaglia and Bray, 1964) with required uniform deviates generated by using a linear congruential generator.

The empirical variances were obtained by a location swindle (Simon, 1976), as follows. The Cauchy variates were $X = Z/Y$, where Z and Y are independent N(0,1). An empirical estimate of σ^2 would be $\tilde{\sigma}^2 = \sum_1^N T^2(\underset{\sim}{x}_j)$, where N is the number of samples and $\underset{\sim}{x}_j$ is the jth sample of Cauchy variates. Note that Var $\tilde{\sigma}^2 =$ (Var $T^2(\underset{\sim}{x})$)/n.

The estimator were made more precise by using a conditioning argument on a translation of the X's. Let $a = \sum_1^{15} Z_iY_i/\sum_1^{15} Y_i^2$. Then

$$\tilde{\sigma}^2 = ET^2(\underset{\sim}{x})$$

$$= E(T(\underset{\sim}{x}-a) + a)^2 \qquad \text{(by location invariance)}$$

$$= E_{\underset{\sim}{x}-a,Y}\left(T^2(\underset{\sim}{x}-a) + \frac{1}{Y'Y}\right) \qquad \text{(by iterated expectation)}$$

$$= E_{\underset{\sim}{x}-a}\, T^2(\underset{\sim}{x}-a) + \frac{1}{13}, \qquad \text{(since } \underset{\sim}{Y}'\underset{\sim}{Y} \text{ is distributed as } \chi^2_{15}\text{)}.$$

Therefore, we can get another unbiased estimate of σ^2 as

$$\tilde{\tilde{\sigma}}^2 = \sum_{1}^{N} T^2(\underset{\sim}{x}_i - a_i)/N + 1/13$$

which has a variance Var $T^2(\underset{\sim}{x}-a)/N$. The increased precision follows from:

$$\text{Var } T^2 = E(\text{Var}(T^2|\underset{\sim}{x}-a,\underset{\sim}{Y})) + \text{Var}(T^2(x-a) + 1/13) \geq \text{Var}(T^2(\underset{\sim}{x}-a)).$$

References

Efron, B. (1975). "Defining the Curvature of a Statistical Problem (with Applications to Second Order Efficiency)," Annals of Statistics, 3, 1189-1217.

Haldane, J.B.S. and S.M. Smith (1956). "The Sampling Distribution of a Maximum Likelihood Estimate," Biometrika, 43, 96-103.

Marsaglia, G. and T.A. Bray (1964). "A Convenient Method for Generating Normal Variables," Siam Review, 6, 260-264.

Rao, C.R. (1963). "Criteria of Estimation in Large Samples," Sankhya, 25, 189-206.

Simon, G. (1976). "Computer Simulation Swindles, with Applications to Estimates of Location and Dispersion," Applied Statistics, 25, 266-274.

FISHER'S DEVELOPMENT OF CONDITIONAL INFERENCE

David Hinkley

1. Introduction

In previous lectures we have examined aspects of Fisher's theory of statistical estimation, where much attention was given to sufficiency, efficiency, and information. In essence the theory is a likelihood-based theory, proposed as an alternative to the then-popular Bayesian theory of Laplace.

One intriguing aspect of the 1925 paper [CP 18] is the discovery that, in the notation of earlier lectures, the maximum likelihood estimator, $\hat{\theta}$, together with the observed information, $I = -\ddot{\ell}_{\hat{\theta}}$, contain all but an asymptotically-vanishing amount of information. Fisher refers to I as "the true weight" for $\hat{\theta}$, as opposed to $E(-\ddot{\ell}_{\theta})$, and coins the term "ancillary statistic" for I. This is the beginning of a conditional theory of inference, which crystallizes in 1934, as we shall shortly see.

The general point of conditional inference is that, inasmuch as inference proceeds by relating a given experimental outcome to a series of hypothetical repetitions which generates frequency distributions for statistics, that series of repetitions should be as relevant as possible to the data at hand. In effect, conditional inference proceeds by conditioning sampling distributions to relevant subsets of the general experimental sample space. The following simple example illustrates this.

Example 1: The unknown weight of an object is θ, which can be measured with error by $x = \theta + e$. There are two available measuring instruments for which errors of measurement have frequency distributions, $N(0,1)$ and $N(0,0.01)$, respectively. Our experiment consists of choosing one of these instruments at random, and then taking one measurement x. In the full sample space for the experiment the estimate x of θ has a mixture distribution

$$\text{pr}(x-\theta \le c) = \tfrac{1}{2}\Phi(c) + \tfrac{1}{2}\Phi(10c),$$

according to which $x \pm 1.645$ is a 95% confidence interval for θ. However, if the second instrument is used, we would in fact use our knowledge that the more precise type of measurement is made, and so obtain the conditional 95% confidence interval $x \pm 0.196$ derived from the conditional $N(0,0.01)$ distribution of $x-\theta$. It would seem unwise to interpret x with partial reference to the behavior of imprecise measurements that were in fact not made.

A somewhat more complicated, but more familiar, version of this example would be that of fitting a linear regression $y = a + bx + e$ to bivariate normal data, where confidence intervals for b, say, are obtained conditional on the experimental values of x.

2. Information Recovery Through Conditioning

Fisher's 1925 reference to $I = -\ddot{\ell}_{\hat{\theta}}$ as a true weight (inverse variance) for $\hat{\theta}$ is somewhat mysterious, and slightly inaccurate. The issue is much clearer in 1934 [CP 108], when Fisher demonstrates a conditional method of inference for the measurement-error model. Here $x_1,\ldots,x_n$ are independent measurements, $x_j = \theta + e_j$, where the e_j are effectively sampled from a frequency distribution with density $f(e)$, say. What Fisher has recognized is that: (i) in general the total sample information is not retained by the MLE $\hat{\theta}$, even though the MLE retains more information than any other estimator; (ii) the pair $(\hat{\theta},I)$ retains more information than $\hat{\theta}$ alone; and (iii) I, which is location invariant in the measurement-error model, of itself contains no information. The implications are that in forming the statistical estimate of θ, no information is available except that in $\hat{\theta}$, and that in inference about θ from the estimate, I provides information. In fact I is but one of several pieces of information which can be used to interpret $\hat{\theta}$, and in the measurement-error problem a complete recovery of information is possible, as the 1934 paper shows.

Consider the sufficient ordered data $x_{(1)} \le \cdots \le x_{(n)}$ in the equivalent form $(\hat{\theta},a_1,\ldots,a_n)$ with a_j defined to be the invariant residuals $x_j-\hat{\theta}$, $j = 1,\ldots,n$; in fact a_n is redundant. The invariants $a_1,\ldots,a_n$, which Fisher terms the configuration of the sample, have joint distribution $g(\underset{\sim}{a})$ which is independent of θ, and hence are termed ancillary statistics. The joint frequency function of the sufficient statistic, $f_\theta^*(x_{(\cdot)}) = n!\,\Pi\, f(x_j - \theta)$, is expressed as

$$f_\theta^*(x_{(\cdot)}) = h_\theta(\hat{\theta}|\underset{\sim}{a})g(\underset{\sim}{a}), \tag{1}$$

and the score function is therefore

$$\dot{\ell}_\theta(\underset{\sim}{x}) = \dot{\ell}_\theta(\hat{\theta}|\underset{\sim}{a}) = \partial \log h_\theta(\hat{\theta}|\underset{\sim}{a})/\partial\theta.$$

The mean sample information $\mathcal{J}_\theta^S$, defined either as $E\{-\ddot{\ell}_\theta(\underset{\sim}{x})\}$ or as Var $\{\dot{\ell}_\theta(\underset{\sim}{x})\}$, is easily seen to satisfy

$$\mathcal{J}_\theta^S = E_{\underset{\sim}{a}}\, \mathcal{J}_\theta^{\hat{\theta}|\underset{\sim}{a}}, \tag{2}$$

where $\mathcal{J}_\theta^{\hat{\theta}|\underset{\sim}{a}}$ is the mean information in the conditional distribution of $\hat{\theta}$ given $\underset{\sim}{a}$. Therefore this conditional distribution is sufficient, in the sense that all information about θ is retained in repeated sampling. The form of the conditional distribution is particularly simple, since by location invariance

$$g(\underset{\sim}{a}) = \int f_\theta^*(x_{(\cdot)})d\hat{\theta} = \int f_0^*(u - \theta + a_1, \ldots, u - \theta + a_n)d(u - \theta)$$

$$= \int f_t^*(x_{(\cdot)})dt = n! \int \Pi\, f(x_j - t)dt,$$

which implies that, in expression (1),

$$h_\theta(\hat{\theta}|\underset{\sim}{a}) = \frac{\text{lik}(\theta|\underset{\sim}{x})}{\int \text{lik}(t|\underset{\sim}{x})dt}, \tag{3}$$

with $\text{lik}(\theta|\underset{\sim}{x}) = \Pi\, f(x_j - \theta)$. Note that $x_i = \hat{\theta} + a_i$ on the right hand side of expression (3). The conditional distribution is of location form, the frequency function for $t = \hat{\theta} - \theta$ being

$$\text{lik}(-t|\underset{\sim}{x}) / \int \text{lik}(u|\underset{\sim}{x})du.$$

Thus the likelihood itself, normalized and reflected about its mode, gives the fully informative distribution for inference about θ.

Fisher is reservedly enthusiastic about this remarkable result: "This ... will seldom be so convenient as the use of an estimate by itself, without reference to the configuration. ... The reduction of the data is sacrificed to its complete interpretation." It is conceivable that Fisher was unaware of the possible large effects of conditioning, a point we shall return to shortly.

From the detailed calculation in the case $f(e) = \frac{1}{2}\exp(-|e|)$, Fisher notes that the local behavior of the likelihood function near $\theta = \hat{\theta}$ is most important. More generally he stresses the second differential, $I = -\ddot{\ell}_{\hat{\theta}}$, as the primary component; the configuration, $a_1,\ldots,a_n$, could be re-expressed in terms of n-1 successive derivatives of log likelihood at $\hat{\theta}$.

The conditional inference obtained restricts comparison of a given sample to hypothetical samples with the same configuration, but of course it relies heavily on the form of the frequency curve assumed. In fact the conditional distribution is mathematically equivalent to a Bayes posterior when uniform prior measure is given to θ.

A very simple example of the theory, first appearing in Fisher's 1956 book [SMSI], is the following.

<u>Example 2</u>: Suppose that $\underset{\sim}{X}_1,\ldots,\underset{\sim}{X}_n$ are independently drawn from the bivariate normal distribution with mean vector $\underset{\sim}{\mu}$, variances equal to one, and zero correlation. Suppose further that μ lies on a known circle of radius ρ, centered at the origin, and write $\mu' = (\rho\cos\theta, \rho\sin\theta)$. In this case $\underset{\sim}{\bar{X}}$ is minimal sufficient and the configuration is the single statistic $r = \|\underset{\sim}{\bar{X}}\|$, the radius of the circle passing through $\underset{\sim}{\bar{X}}$. A simple calculation shows that

$$f_\theta(\hat{\theta}|r) = \frac{1}{2\pi I_0(nr\rho)}\exp\{nr\rho\cos(\hat{\theta}-\theta)\}.$$

The conditioning can be illustrated geometrically, as in Figure 1, where two different sample values, $\underset{\sim}{\bar{X}}$ and $\underset{\sim}{\bar{X}}^*$, are surrounded by standard 95% confidence regions for $\underset{\sim}{\mu}$. The larger is r, the more precisely is the angle θ located by the tangent rays to the confidence circle; the equivalent confidence arcs for μ, on the circle $\|\mu\| = \rho$, have end-points where those tangent rays cut $\|\mu\| = \rho$.

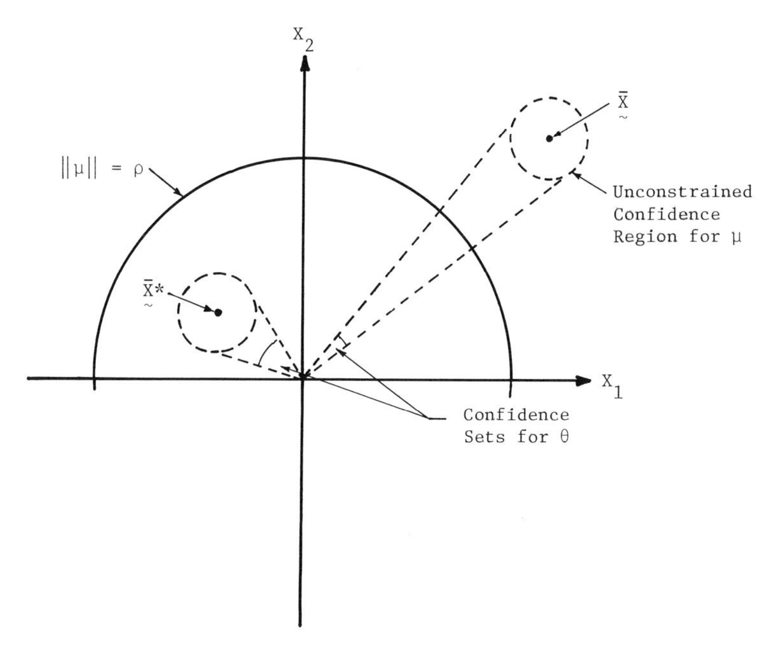

Figure 1. Two Different Conditional Confidence Sets for θ in Fisher's Normal Circle Model, Derived from Unconstrained Confidence Sets for μ.

3. The General One-Parameter Case

The exact conditional solution for the measurement error model, which Fisher also gave for the case of unknown error scale, becomes the fiducial distribution for θ by inverting the roles of θ and $\hat{\theta}$. Somewhat lower status was given to cases where exact ancillary statistics do not exist, but nevertheless Fisher clearly recognized the general principle of conditioning in reference to the standard error for $\hat{\theta}$. Thus in his 1925 book [SMRW], in §57.2, he refers to $I = -\ddot{\ell}_{\hat{\theta}}$ in an example as follows: "It should be noted that an estimate [namely I, an estimate of $\mathcal{I}_\theta$] obtained thus is in no way inferior to one [$\mathcal{I}_\theta$] obtained from theoretical expectations. ... It might be said that owing to chance the experiment has given a somewhat higher amount of information than should be expected ...". There is then reference to the distinction between $\mathcal{I}_\theta$ and I as being of theoretical interest.

The point about using I as inverse variance is clear in the case of Example 2, since by expanding $\cos(\hat{\theta}-\theta)$ in power series we find that

$$f_\theta(\hat{\theta}|r) = \text{constant} \times \exp\{-\tfrac{1}{2}\, nr\rho(\hat{\theta}-\theta)^2 + O((\hat{\theta}-\theta)^4)\},$$

giving an approximate normal curve with weight $I = nr\rho$. This approximation is very accurate when $nr\rho \geq 4$.

In the general location model the conditional distribution is not so simple, and a normal approximation to $f_\theta(\hat{\theta}|\underset{\sim}{a})$ given by expression (2) ignores some information. Nevertheless, one can show that, conditional on $\underset{\sim}{a}$, the χ_1^2 approximation is appropriate for $I(\hat{\theta}-\theta)^2$ but not in general for $\mathcal{J}(\hat{\theta}-\theta)^2$. Moreover, the approximation is even better for $2(\ell_{\hat{\theta}}-\ell_\theta)$. That this is so has been demonstrated by Efron and Hinkley (1978), who found convincing evidence for applicability of these results beyond the exact case of measurement-error models. A dramatic illustration is provided by the Cauchy error distribution, where the coefficient of variation of I is $\sqrt{5/2n}$. Figure 2 gives reliable Monte Carlo estimates of probabilities with which $2(\ell_{\hat{\theta}}-\ell_\theta)$, $I(\hat{\theta}-\theta)^2$, and $\mathcal{J}(\hat{\theta}-\theta)^2$ exceed 3.84 (nominal 95% point for χ_1^2) in samples of size 20, grouped by interval values of I. Similar effects have been found for other models.

4. Difficulties with Conditional Inference

In Fisher's 1934 invited address to the Royal Statistical Society [CP 124] -- which was not a friendly encounter, judging by the discussion -- Fisher introduced conditional inference with reference to the "well-worn topic of the 2×2 table". Neither set of margins is fixed *a priori*, although it is natural to think of the table as a comparison of two binomial proportions with row totals fixed. Fisher then states: "Let us blot out the contents of the table, leaving only the marginal frequencies. If it be admitted that these marginal frequencies by themselves supply no information ... as to the proportionality of the frequencies ..., we may recognize the information they supply as wholly ancillary...". The parameter of interest here is the log odds ratio, and the joint distribution of the margins is *not* independent of this parameter, albeit nearly so (Plackett, 1975).

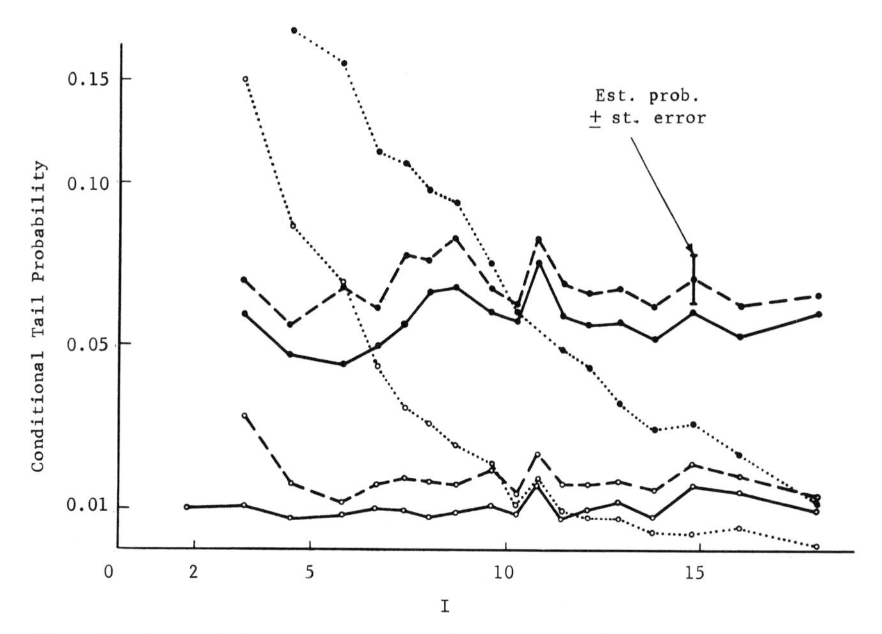

Figure 2. Monte Carlo Estimates of pr (statistic $\geq$ c|I) with c = 3.84, shown by closed circles, and c = 5.99, open circles, for three likelihood statistics in the Cauchy location model, n = 20. Statistics: $2(\ell_{\hat{\theta}}-\ell_{\theta})$ shown by solid curve; $I(\hat{\theta}-\theta)^2$, dashed curve, $\mathcal{J}(\hat{\theta}-\theta)^2$, dotted curve.

Note that the corresponding exact test of equal proportions does condition on both sets of margins, and this test is widely accepted, in part because of the mathematical property of similarity (Cox and Hinkley, 1974, p. 134). Fisher's use of conditioning for estimation is not widely accepted, but it has led to discussion of more general definitions of ancillarity (Barndorff-Nielsen, 1978).

In the general context of inferential theory, conditioning and ancillarity encounter difficulties of various kinds. Without going into detail, these difficulties include (i) non-uniqueness of ancillary statistics, (ii) a purported proof that acceptance of conditioning as a general principle implies that only the form of the observed likelihood function should be used for inference (the so-called likelihood principle), (iii) the claim that conditioning reduces the power of tests of hypotheses. The latter claim seems to me essentially void, since it rests on the notion of fixed significance levels, and in any event the prime requisite of any inference is that it be relevant.

One interesting aspect of conditioning is the distinction between experimental and mathematical ancillaries. The former correspond to definitions of achievable experiments, as in Example 1; the latter correspond to mathematical artifacts, as in Fisher's measurement-error model. A detailed discussion is given by Kalbfleisch (1975).

Finally, notice that an exact ancillary statistic provides the basis for a test of the adequacy of the assumed model. As one can see in the measurement-error model, the most dramatic differences between conditional and unconditional inferences occur when the assumed model is to some extent contradicted by the ancillary statistic. This point deserves some attention.

References

Barndorff-Nielsen, O. (1978). Information and Exponential Families in Statistical Theory. New York: John Wiley & Sons.

Cox, D.R. and D.V. Hinkley (1974). Theoretical Statistics. London: Chapman and Hall.

Efron, B. and D.V. Hinkley (1978). "Assessing the Accuracy of the Maximum Likelihood Estimator: Observed Versus Expected Fisher Information," Biometrika, 65, 457-482.

Kalbfleisch, J.D. (1975). "Sufficiency and Conditionality (with Discussion)," Biometrika, 62, 251-268.

Plackett, R.L. (1977). "The Marginal Totals of a 2×2 Table," Biometrika, 64, 37-42.

This research was supported by National Science Foundation Grant MCS 7904558 to the University of Minnesota.

FIDUCIAL INFERENCE

Robert Buehler

1. Introduction

The concept of fiducial probability is due to R.A. Fisher. The fiducial distribution of a parameter θ given an observation x is intended to describe our uncertainty about θ after x is observed when there was no *a priori* information about θ. The term "fiducial distribution" first appeared in Fisher's 1930 paper [CP 84], where the distribution of the correlation coefficient was considered. Unfortunately, Fisher never gave an acceptable general definition of fiducial probability. For the case of one observation x and one parameter θ, with cumulative sampling distribution $F(x|\theta)$ monotone decreasing in θ, Fisher defined the fiducial density for θ to be

$$f(\theta|x) = -\frac{\partial F(x|\theta)}{\partial\theta}. \tag{1}$$

In this presentation, we will review some of Fisher's work on fiducial inference.

2. Geometrical Construction of a Fiducial Distribution

Suppose that X is distributed $N(\theta,1)$, and that we are given any sequence of θ-values, $\theta_1,\theta_2,\ldots$. Let x_1 denote an observation from $N(\theta_1,1)$, x_2 from $N(\theta_2,1)$, etc. In Figure 1, the values $(\theta_1,x_1),(\theta_2,x_2),\ldots$ are plotted on the left-hand scale. A single point is designated x, and θ-values are plotted on the right-hand side, relative to x in such a way that θ_i-x on the right equals θ_i-x_i on the left. In the long run, we get a collection of θ-values on the right-hand side which behaves exactly like a random sample from the distribution $N(x,1)$. The same construction can be used for any distribution such that θ is a location parameter for x, and the fiducial density of θ given x is found to be the mirror image of the density of x given θ. Thus when we observe x_1 our uncertainty about θ_1 is equivalent to the knowledge that it is a random value from the $N(x_1,1)$ frequency distribution.

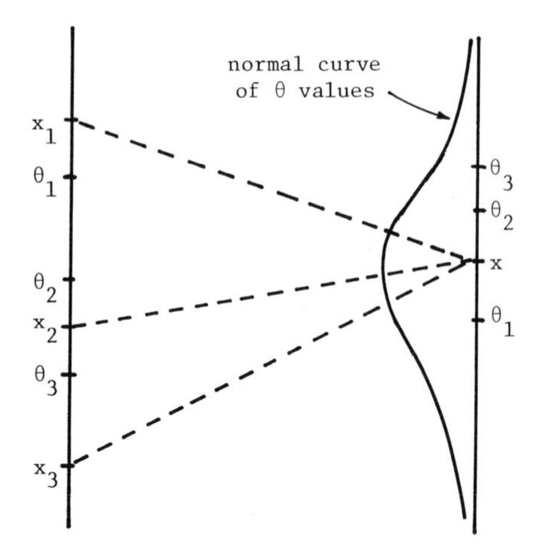

Figure 1: Geometrical Interpretation of Fiducial Distribution

3. The 1930 Fiducial Paper [CP 84]

Suppose that T is a statistic computed from a random sample and that θ is the corresponding population parameter. Then, according to Fisher:

> There is a relationship between T and θ such that T is the 95 percent value corresponding to a given θ, and this relationship implies the perfectly objective fact that in 5 percent of samples T will exceed the 95 percent value corresponding to the actual value of θ in the population from which it is drawn. To any value of T there will moreover be usually a particular value of θ to which it bears this relationship; we may call this the "fiducial 5 percent value of θ" corresponding to a given T.

Fisher gives as an illustrative example the case of the sample correlation r computed from four pairs of observations with ρ the corresponding population correlation. A table is included in the paper to give the 95 percent value of r given ρ and the 5 percent fiducial point of ρ given r; in other words, pairs of values r,ρ such that $F(r|\rho) = 0.95$.

4. The 1934 Paper [CP 108]

Here Fisher considers the fiducial distribution of the shape parameter θ in the gamma density $f(x,\theta) = \frac{1}{\theta!} x^{\theta} e^{-x}$. The exact fiducial distribution of θ is expressed as an integral, and limiting forms for small and large values of θ (θ near -1 and near ∞) are given. The paper goes on to generalize the gamma distribution theory by considering what we now would call an exponential family. Location families are then taken up and illustrated by detailed calculations for the double exponential (Laplace) distribution. Finally location and scale parameters are considered jointly. The well-known ancillary or configuration statistics appear here. In this work, the emphasis is on the structure of the likelihood function, and on the amount of information in the sample and in the maximum likelihood estimator (see also Hinkley's lecture on conditional inference). Somewhat surprisingly, the fiducial distributions of location and scale parameters are not given explicitly, or even mentioned. The analysis is, however, relevant to these, and the explicit expressions were given later by Pitman (1939).

5. The 1935 Paper [CP 125]

Suppose a sample of n observations, $x_1,\ldots,x_n$, is drawn from a normal population with mean μ, then we know that $t = \sqrt{n}\,(\bar{x} - \mu)/s$ has a "Student" t-distribution with n-1 degrees of freedom, where $\bar{x} = \Sigma\, x/n$ and $s^2 = \Sigma(x - \bar{x})^2/(n-1)$. Therefore, for any percentile of this t-distribution, say, t_p, the following two inequalities will be satisfied with the same probability:

$$t > t_p, \qquad (2)$$

$$\mu < \bar{x} - st_p/\sqrt{n}\,. \qquad (3)$$

By varying t_p, the right-hand side of expression (3) will take all real values. Thus

> ... we may state the probability that μ is less than any assigned value, ..., or, in short, its probability distribution, in the light of the sample observed.... The distribution which we have obtained is independent of all prior knowledge of the distribution of μ, and is true of the aggregate of all samples without selection. It involves $\bar{x}$ and s as parameters, but does not apply to any special selection of these quantities. To distinguish it from any of the inverse probability distributions derivable from the same data it has been termed the fiducial probability distribution, and the probability statements which it embraces are termed statements of fiducial probability.

If we use s', derived from the mean (absolute) error, instead of s, then the quantity $t' = \sqrt{n}\,(\bar{x} - \mu)/s$ is also pivotal; that is, its distribution depends only on the sample size and not on any unknown parameters. Therefore, any probability statement for t' could also be expressed in terms of μ. Fisher hastens to point out that the use of s' in place of s does not give a valid fiducial distribution because s' is not sufficient and so "to use s' in place of s would be logically equivalent to rejecting arbitrarily a portion of the observational data and basing probability statements on the remainder as though it had been the whole".

Fisher goes on to discuss, for the first time, the fiducial distribution of a future observation. Take the normal case, for example, with the same definitions for $\bar{x}$ and s, and let x' denote a future observation. Then $t = (x' - \bar{x})/s$ is ancillary and the probability of $t_{n-1} > t_1$ is the same as $\frac{x'-\bar{x}}{s} > t_1\sqrt{\frac{n+1}{n}}$ with x' the only unknown element. Therefore, we can make probability statements about x'. This problem of prediction is generalized: given a sample of n observations yielding the statistics $\bar{x}$ and s, find the fiducial frequency distribution of the statistics $\bar{x}'$ and s' derived from a subsequent sample of n' observations. If we let $n' \to \infty$, then the statistics $\bar{x}'$ and s' tend to the values μ and σ. The consequent simultaneous fiducial distribution of μ and σ is shown by Fisher to be

$$df = \frac{\sqrt{n}}{\sigma\sqrt{2\pi}}\, e^{-n(\mu-\bar{x})^2/2\sigma^2}\, d\mu\, \frac{1}{\frac{1}{2}(n-3)!}\, \{(n-1)s^2/2\sigma^2\}^{\frac{1}{2}(n-1)}\, e^{-(n-1)s^2/2\sigma^2}\, \frac{2d\sigma}{\sigma} .$$

The marginal distributions of μ and σ, derived by integrating this joint distribution with respect to the other parameter, are in accordance with the facts that $\sqrt{n}\,(\mu-\bar{x})/s$ is distributed as t_{n-1} and $(n-1)s^2/\sigma^2$ as χ^2_{n-1}.

With these results in mind, Fisher argues that:

> In general, it appears that if statistics $T_1, T_2, T_3, \ldots$ contain jointly the whole of the information available respecting parameters $\theta_1, \theta_2, \theta_3, \ldots$ and if functions $t_1, t_2, t_3, \ldots$ of the T's and θ's can be found, the simultaneous distribution of which is independent of $\theta_1, \theta_2, \theta_3, \ldots$, then the fiducial distribution of $\theta_1, \theta_2, \theta_3, \ldots$ simultaneously may be found by substitution....

This rather tentative assertion prompted later researchers to find counterexamples showing that joint pivotals need not lead to unique fiducial distributions.

Another case discussed is the difference between the means of two normally distributed populations. Let a sample of n observations yield a mean $\bar{x}$, and an estimated variance of the mean s^2, i.e., $s^2 = \Sigma(x-\bar{x})^2/n(n-1)$. Then from previous results we know $\mu = \bar{x} + st$ where t has the "Student" t-distribution with n-1 degrees of freedom. Similarly, for the mean of a second population, of which we have n' observations, $\mu' = \bar{x}' + s't'$, where t' has n'-1 degrees of freedom. If we write $\mu' - \mu = \delta$, $\bar{x}' - \bar{x} = d$, then $\varepsilon = \delta - d = s't' - st$ is a weighted sum of two "Student" variates and has the Behrens distribution (Behrens, 1929), which Fisher later assisted in tabulating. [Some related material is found in CP 151, CP 162, CP 264, and CP 268.] See also the lecture by Wallace on the Behrens-Fisher problem.

The last part of the paper deals with "the variance of a normally distributed set of means". Suppose we have k samples of n observations from equally variable normal populations, i.e., the model under consideration is

$$y_{ij} = \mu + a_i + e_{ij}, \qquad (i = 1,\ldots,k;\ j = 1,\ldots,n),$$

where $\mathrm{Var}(a_i) = \phi$, $\mathrm{Var}(e_{ij}) = \theta$. The parameter of interest is ϕ. The usual analysis of variance table is

	df	SS	Mean Square
Among Samples	k-1	$n\sum_i(\bar{y}_{i.}-\bar{y}_{..})^2$	a
Error	k(n-1)	$\sum_i\sum_j(y_{ij}-\bar{y}_{i.})^2$	b

where

$$a = \frac{\chi_1^2}{n_1}(\theta + n\phi)$$

and

$$b = \frac{\chi_2^2}{n_2}\theta,$$

with $n_1 = k-1$ and $n_2 = k(n-1)$. Hence

$$a \frac{n_1}{\chi_1^2} - b \frac{n_2}{\chi_2^2} = n\phi.$$

Fisher gives $\theta + k\phi$ instead of $\theta + n\phi$, apparently an error.

We wish to obtain the distribution of ϕ. Since χ_1^2, χ_2^2 are independent, the distribution of ϕ may be calculated from their simultaneous distribution. In Figure 2, if n_1/χ_1^2 and n_2/χ_2^2 are the coordinates of a point, then the points consistent with any given value of ϕ lie on a straight line, making an angle with the axis of n_1/χ_1^2, the tangent of which is a/b. The fiducial probability of ϕ exceeding any chosen value is the total frequency to the right of the corresponding line.

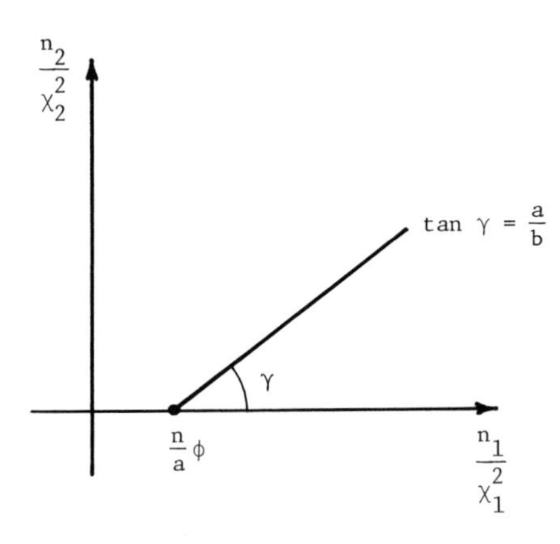

Figure 2: Construction of Fiducial Distribution for Variance Component

One interesting question is "What is the fiducial probability of $\phi < 0$?", since lines which do not strike the axis of the abscissa to the right of the origin correspond to negative values of ϕ. Fisher did not answer this question; he simply stated that this problem was of no interest. This is true if we want only to test the hypothesis $\phi = 0$. Nevertheless the logical question of where to assign the fiducial probability associated with negative ϕ is one which has puzzled later researchers.

6. The 1936 Paper [CP 137]

Here Fisher states the "Problem of the Nile":

> The agriculturan land of a pre-dynastic Egyptian village is of unequal fertility. Given the height to which the Nile will rise, the fertility of every portion of it is known with exactitude. But the height of the flood affects different parts of the territory unequally. It is required to divide the area between several households of the village, so that the yields of the lots assigned to each shall be in pre-determined proportion, whatever may be the height to which the river rises.

The problem in mathematical terms is as follows: Let $f(x_1,x_2,\theta)$ be the fertility at (x_1,x_2) given height of flood θ, and write $g(x_1,x_2,\theta) = f(x_1,x_2,\theta)/\iint_\Omega f(x_1,x_2,\theta)dx_1dx_2$ where Ω is the total area. Denote the ith plot by A_i. Then $\iint_{A_i} fdx$ is the absolute yield of plot A_i and $\iint_{A_i} gdx$ is the relative yield of plot A_i, and moreover $\iint_{A_i} gdx$ must not depend on θ. Therefore, the indicator function of A_i is an ancillary statistic, and if the boundaries between the A_i can be defined by arbitrary constant values of a continuous statistic, then that statistic is ancillary. An unsolved problem in 1936, and still unsolved today, is: Given a particular $g(x_1,x_2,\theta)$, does there exist an ancillary statistic?

7. The 1939 Paper [CP 162]

In this paper Fisher defends Behrens' test, referred to in Section 5. In referring to this paper later in the volume of collected papers, Fisher wrote "Pearson and Neyman had laid it down axiomatically that the level of the significance of a test must be equated to the frequency of a wrong decision 'in repeated samples from the same population'. This idea was foreign to the development of tests of significance given by the author in 1925". Here Fisher appears to depart from his earlier frequentist views; see the lectures by Lane and Wallace in this volume.

8. The 1939 Paper [CP 165]

Fisher describes the fiducial distribution of percentiles given the order statistics $x_{(1)},\ldots,x_{(n)}$ as follows. "Let μ_p stand for the hypothetical value which exceeds the fraction p of the population sampled, then the terms of the expansion of $(q+p)^n$, $p+q = 1$, will give in succession the probabilities that μ_p is less than all the observed values, exceeds one of them, exceeds two, or finally exceeds them all." Thus when we have only two observations, $x_{(1)}$ and

$x_{(2)}$, and μ_p is the median of the population, for example, the general rule gives the fiducial probabilities ¼, ½, ¼ for the events $\mu_p \leq x_{(1)}$, $x_{(1)} \leq \mu_p \leq x_{(2)}$, $x_{(2)} < \mu_p$.

Fisher also deduced the joint distribution of the percentiles given the order statistic $x_{(1)},\ldots,x_{(n)}$. If $p_1,\ldots,p_n$ denote the unknown parameters $F(x_{(1)}),\ldots,F(x_{(n)})$, then $(p_1,\ldots,p_n)$ has a uniform distribution on the simplex $\{(p_1,\ldots,p_n) | 0 < p_1 < \ldots < p_n < 1\}$.

9. The 1956 Book, Statistical Methods and Scientific Inference [SMSI]

I will mention only a few points from this book which contains some of Fisher's final attempts to explain fiducial probability.

(1) Fisher again emphasized the difference between the fiducial argument and the Bayesian approach: the latter requires a distribution *a priori* of the parameter while the former uses the observations only to change the logical status of the parameter from one in which nothing is known of it, and no probability statement about it can be made, to the status of a random variable having a well-defined distribution. (Again, see the lecture by Lane in this volume).

(2) In his 1935 paper [CP 125], Fisher had derived the joint distribution of μ and σ^2 as

$$\sqrt{\frac{N}{2\pi\sigma^2}}\; e^{-N(\mu-\bar{x})^2/2\sigma^2}\, d\mu \;\frac{1}{\left(\frac{N-3}{2}\right)!}\left(\frac{s}{2\sigma^2}\right)^{\frac{1}{2}(N-1)} e^{-s/2\sigma^2}\,\frac{d\sigma^2}{\sigma^2}.$$

Here, Fisher points out that the rigorous step-by-step demonstration by the fiducial argument would in fact consist of, first, the establishment of the second factor giving the distribution of σ given s, disregarding the other parameter μ, and then the derivation of the first factor as the distribution of μ given $\bar{x}$ and σ.

(3) An example of the "Problem of the Nile" is given. Let the joint density of x and y be $f(x,y|\theta) = \exp(-\theta x - y/\theta)$, and for a sample of n pairs let $X = \Sigma x_i$, $Y = \Sigma y_i$. Then the MLE is $\hat{\theta} = Y/X$, and $Y = XY$ is ancillary. The problem of fiducial limits for θ could be handled by the transformation $z_1 = \log x_1,\ldots,z_n = \log x_n$, $z_{n+1} = \log y_1,\ldots,z_{2n} = \log y_n$, $\phi = \log\theta$. Then ϕ is a location parameter, and its fiducial density is proportional to the likelihood function.

(4) The concept of "recognizable subset" is heavily emphasized (see, for example, pages 57, 80, 109). In connection with the fiducial distribution based on "Student's" t-distribution, Fisher remarks that "since $\bar{x}$ and s are jointly sufficient for estimation, and knowledge of μ and σ *a priori* is absent, there is no possibility of recognizing any subset of cases within the general set, for which any different value of the probability should hold".

10. Related Work

Fiducial inference became a controversial subject as soon as others realized that its results differed from the confidence level results of Neyman. The Behrens problem was a focal point for the early discussion (see CP 264 and references given there). Another controversial example was that of the ratio of two normal means (Fieller, 1954 and Creasy, 1954 with discussion by Fisher reproduced in CP 257). The Behrens and Fieller-Creasy problems are discussed in detail in the lecture by Wallace in this volume.

A difficulty with the notion of relevant subsets was pointed out by Buehler and Feddersen (1963), who provided a counterexample to Fisher's claim concerning non-existence of relevant subsets in "Student's" problem. For some other interesting criticisms of fiducial inference see Anscombe (1957), Tukey (1957), Lindley (1958), and Brillinger (1962).

A direct descendant of fiducial inference based on pivotal statistics is structural inference, developed by Fraser (1968). Finally, Wilkinson (1977) has recently made a bold attempt to rejuvenate fiducial inference.

References

Anscombe, F.J. (1957). "Dependence of the Fiducial Argument on the Sampling Rule," *Biometrika*, 44, 464-469.

Behrens, W.-V. (1929). "Ein Betrag zur Fehlenberechnung bei wenigen Beobachtungen," *Landwirtschaftliche Jahrbücher*, 68, 807-837.

Buehler, R.J. and A.P. Feddersen (1963). "Note on a Conditional Property of Student's t," *The Annals of Mathematical Statistics*, 34, 1098-1100.

Brillinger, D.R. (1962). "Examples of the Definition of Fiducial Probability with a Bibliography," *The Annals of Mathematical Statistics*, 33, 1349-1355.

Creasy, M.A. (1954). "Limits for the Ratio of Means," *Journal of the Royal Statistical Society, Series B*, 16, 186-194.

Fieller, E.C. (1954), "Some Problems in Interval Estimation," *Journal of the Royal Statistical Society, Series B*, 16, 175-185.

Fraser, D.A.S. (1968). *The Structure of Inference*. New York: John Wiley and Sons.

Lindley, D.V. (1958). "Fiducial Distributions and Bayes' Theorem," *Journal of the Royal Statistical Society*, Series B, 20, 102-107.

Pitman, E.J.G. (1939). "The Estimation of the Location and Scale Parameters of a Continuous Population of any Given Form," *Biometrika*, 30, 391-421.

Tukey, J.W. (1957). "Some Examples with Fiducial Relevance," *The Annals of Mathematical Statistics*, 28, 687-695.

Wilkinson, G.N. (1977). "On Resolving the Controversy in Statistical Inference," *Journal of the Royal Statistical Society, Series B*, 39, 119-171.

THE BEHRENS-FISHER AND FIELLER-CREASY PROBLEMS

David L. Wallace

1. Introduction

Of all R.A. Fisher's creations, fiducial inference was perhaps the most ambitious, yet least accepted. Ignored by a large part of the statistics community satisfied with the mathematically simpler confidence approach, and rejected as logically imperfect and inconsistent in general by those who recognized the strength of the fiducial objectives, the fiducial argument continues under active and sympathetic study today only in a few islands of the statistical world.

I assume that my audience is composed mostly of those to whom the confidence argument is standard, to whom the Bayesian argument is seen as a straightforward computation from possibly unjustifiable prior assumptions, and for whom the fiducial argument may seem artificial or perhaps indistinguishable from the Bayesian argument. I examine only one aspect of the fiducial argument and only in the context of some two-sample problems. I hope to show that the objectives and ideas of the fiducial argument in that context are attractive and stand in contrast to the confidence argument.

Inference about a normal mean started modern theory and the practice of small sample inference. "Student's" work is accepted and recognized as basic by all. Setting aside doubts on the Gaussian and independence assumptions, we have wide acceptance for inferential statements in the form of 95% limits (and of other levels) on the unknown mean μ_1 even if we cannot agree on the adjective modifying "limits". The uncertainty about μ_1 is conveniently and at least schematically represented by a t-distribution centered at the observed mean and scaled by the estimated standard error of the mean, with limits given by the appropriate fractiles, as illustrated for each margin in Figure 1.

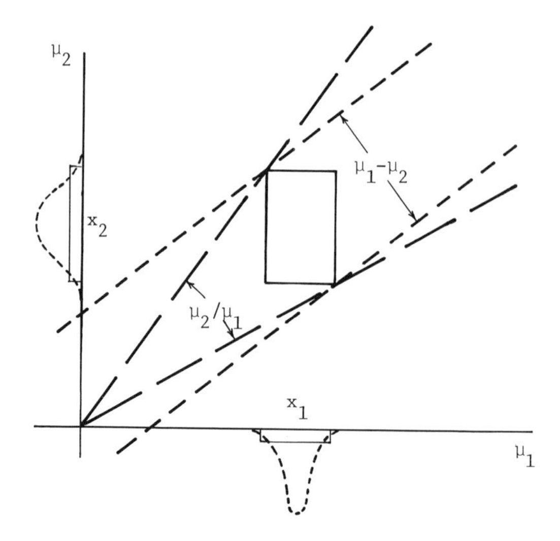

Figure 1. Inferential Distributions and Limits; Conservative Implied Limits on Differences and Ratios.

If we have independent samples from two normal populations, we can separately or jointly set limits on the two means. The Behrens-Fisher and Fieller-Creasy problems are, respectively, the problems of making exact inferences about the difference $\mu_1 - \mu_2$ and the ratio μ_2/μ_1. A joint inferential region for (μ_1,μ_2) evidently implies conservative limits on the difference or the ratio as shown in Figure 1. The problem lies in constructing exact inferences. For the difference $\mu_1-\mu_2$, difficulties arise only when the error variances in the two populations are unknown and in unknown ratio. For the ratio μ_2/μ_1. the problem contains all its essential difficulties even when both error variances are known. Most of this presentation will focus on the Behrens-Fisher problem, mainly because Fisher's ideas and involvement are vastly greater there.

2. Comparing Two Means with Possibly Unequal Variances

Fisher repeatedly stressed the relative unimportance in practice of what has come to be known as the Behrens-Fisher problem. Three quotations will convey some of the spirit and detail. From Statistical Methods and Scientific Inference [SMSI, 1958, p. 94]:

> These early results covered immediate practical requirements rather fully, for with large samples capable of supplying, on internal evidence, accurate estimates of precision, the large-sample procedure of estimating the two sources of error independently could be relied on; and, in a great deal of practical experimental work with small samples, although different lots of material might in reality have somewhat unequal variances, there were good reasons for supposing the real differences to be small compared with the errors of estimation from the small samples individually; so that better comparisons would be obtainable by pooling the variances of the different lots. The mathematical problem of the comparison of means of samples, not only small in size, but for which there is no reason *a priori* to dismiss the largest imaginable differences in precision, was of mathematical interest, and potential experimental importance, though it is difficult to find realistic data which present this problem.

From Statistical Methods for Research Workers [SMRW, Section 24.1]:

> It has been repeatedly stated, perhaps through a misreading of the last paragraph, that our method involves the "assumption" that the two variances are equal. This is an incorrect form of statement; the equality of the variances is a necessary part of the hypothesis to be tested, namely that the two samples are drawn from the same normal population. The validity of the t-test, as a test of this hypothesis, is therefore absolute, and requires no assumption whatever. It would, of course, be legitimate to make a different test of significance appropriate to the question: Might these samples have been drawn from different normal populations having the same mean? This problem has, in fact, been solved, but in relation to the real situations arising in biological research, the question it answers appears to be somewhat academic. Numerical tables of this test were first calculated by W.V. Behrens (1929) and much more completely by P.V. Sukhatmé. These are of use, when there is reason to suspect unequal variances, in removing any doubt from the interpretation of the test of significance.

From his second paper on the topic [CP 162, p. 192]:

> The advances of statistical science have consisted largely in the provision of exact tests of significance appropriate to an increasing variety of useful hypotheses, and occasionally, though not characteristically in experimental work, some interest attaches to hypotheses implying that the means of two populations are equal, while their variances are unequal. At least a theoretical problem of this sort can be framed.

I return to these quotations in the course of considering some examples and issues of principles.

As a first example, I excerpt from Rayleigh's (1893-1894) determinations of the mass of nitrogen, taking the data for determination from one chemical source and from one method using air as the source. The data are shown in Table 1, linearly coded from the original units, along with the usual summary of statistics. With means of 1025 and 8, subject to standard errors of 3.2 and 91 respectively, there is an apparently significant difference in mass as determined by the two methods, as well as substantial difference in the precision of the two methods. The degrees of freedom are quite small. What is an appropriate allowance for error for the estimated difference 1017 between the means?

Table 1: Rayleigh's Determinations of the Mass of Nitrogen. Mass coded to (original units - 2.3) $\times 10^5$.

	From air by hot copper (1893)	From iron and nitrous oxide (1894)
	1035	143
	1026	-110
	1024	-184
	1012	132
	1027	
Means	$\bar{x}_1 = 1025$	$\bar{x}_2 = 8$
Variances	$s_1^2 = 69$	$s_2^2 = 33096$
Standard Errors of Means	$s_1 \div \sqrt{n_1} = 3.71$	$s_2 \div \sqrt{n_2} = 90.96$

Difference $\bar{x}_1 - \bar{x}_2 = 1017$

Standard Error $\sqrt{(s_1^2/n_1 + s_2^2/n_2)} = 91.04$

We have here a clear instance of the Behrens-Fisher problem. The null hypothesis specifies only the equality of the means; the methods are different chemical processes and there is no reason for their variabilities to be similar. The experiment is not a randomized comparison of treatments and Fisher's quoted strictures are not apt.

An ad hoc development of a correct solution for these particular data is easy. The standard error (91.04) for the difference differs from that (90.96) for the chemical source mean only in the fourth significant figure, well below sensible rounding for a standard error. Heuristically, the variability is so much greater for the chemical method that practically all the uncertainty is in one mean. We are effectively making a one-sample comparison of the chemical mean, 8, against the "known" value 1025 from air. The standard error is thus only estimated with 3 degrees of freedom and we would use a critical value of t to give a 95% allowance of $3.18 \times 91.0 = \pm 290$. Both the fiducial and the leading confidence solutions to the Behrens-Fisher problem essentially reduce to this heuristic solution. In contrast, the two-sample t procedure, surely inappropriate here, gives a much smaller allowance of $\pm$ 190 from a pooled standard error of 80.0 and a t value on 7 degrees of freedom.

In this example, apart from avoiding the pooled t solution, there is no practical difference between approaches. But what if the ratio of error variances had been 50 or 5 instead of 500? Although not too common, the problem can arise, and we ought to know how to deal with it.

I now cast the problem into a standard form and provide notation only for the sufficient statistics in a form showing applicability to means, regression coefficients, or whatever. The observed normal data can be reduced to sufficient statistics x_1, x_2, v_1, v_2, all independently distributed, the x_i normally distributed with mean μ_i and variance V_i, and v_i/V_i distributed in a mean square distribution on f_i degrees of freedom. Symbolically, write:

$$x_i \sim N(\mu_i, V_i)$$

$$v_i/V_i \sim MS(f_i).$$

The difference $\eta = \mu_1 - \mu_2$ is the object of interest and the estimated difference $y = x_1 - x_2$ is distributed as $N(\eta, V_1 + V_2)$.

As auxiliary notation, I introduce the sample and population variance ratios

$$r = v_1/v_2 \qquad \rho = V_1/V_2,$$

and, what is equivalent and often preferable to work with,

$$c_i = v_i/(v_1 + v_2) \qquad \gamma_i = V_i/(V_1 + V_2),$$

as the sample and population fractions attributable to the i-th "mean". As long as we have only two variances, $c_2 = 1 - c_1$ and we often write c for c_1. Note that in the usual application to sample means in samples of sizes n_1 and n_2, equal underlying error variances correspond to the ratio $\rho = n_2/n_1$ and the fraction $\gamma_1 = n_2/(n_1 + n_2)$.

For either sample, the random quantity

$$t = \frac{x - \mu}{\sqrt{v}}$$

is a pivotal quantity, distributed in repeated sampling with μ and V fixed as "Student's" t on f degrees of freedom. Three approaches to inference all lead to describing uncertainty about μ by an inferential distribution that is a scaled t, i.e., $\mu = x + \sqrt{v}\,\tilde{t}$. In the confidence approach, without further restrictions or assumptions, nested confidence intervals can be constructed according to the frequency interpretation: $P\{|x - \mu| < t_\varepsilon \sqrt{v} | \mu, V\} = 1 - \varepsilon$. In the fiducial approach, in the absence of information on μ and on V, the pivotal quantity t induces, for the observed x and v, a fiducial distribution on μ as above. In the invariant, improper Bayesian approach, with prior density $P(d\mu dV) \propto d\mu dV/V$, *a posteriori* μ has the scaled t distribution as above.

The effect on the user of the resulting inference is much the same for all three approaches. Did "Student" see clearly which of the three he was introducing? The confidence and the Bayesian approaches amount to deductive mathematical statements that are part of our elementary statistical education. The logic in the fiducial argument is not so simple, resulting in the conditioned result of the invariant Bayesian without the specification of a prior, but with the elusive but essential "in the absence of information on the values of μ and V". Fisher set forth the development in this form in [CP 125] and elsewhere, and the argument has most recently and carefully been analyzed and qualified by Wilkinson (1978). Later, I give Fisher's alternative development that I find more compelling.

With two independent samples and in the absence of information relating (μ_1, V_1) and (μ_2, V_2), the joint fiducial distribution of the two means is the product distribution. Behrens (1929) proceeded formally to find the marginal distribution of the difference $\mu_1 - \mu_2$ and, with attention to the logic, Fisher [CP 125] developed this as the fiducial distribution. Not available in closed form, it is just the distribution of a linear combination of t-variates and, properly scaled, is usefully named the Behrens-Fisher distribution:

$$\tilde{\mu}_1 - \tilde{\mu}_2 = x_1 - x_2 + \sqrt{v_1+v_2}\,(\sqrt{c_1}\,\tilde{t}_1 - \sqrt{c_2}\,\tilde{t}_2)$$

and

$$\frac{\tilde{\eta} - y}{\sqrt{v_1+v_2}} \sim BF(f_1, f_2, \theta = \sin^{-1}\sqrt{c_1}).$$

Jeffreys (1940) noted that with invariant prior distributions independently on each pair of means and variances, the fiducial distribution given above coincides with the invariant Bayesian posterior distribution and he advocated using the Behrens-Fisher solution. Of course, there can be Bayesian solutions based on priors other than the invariant ones.

The foregoing has no direct confidence counterpart. Instead, one seeks a critical function $d(c)$ of the fraction c such that

$$P\{|\tilde{y} - \eta| > d(c)\sqrt{v_1+v_2}\,|\,\mu_1, \mu_2, V_1, V_2\} = \varepsilon$$

for all V_1, V_2, which would then yield a level ε similar test of $\eta = 0$ and lead to exact confidence limits on η. Whether there existed any non-randomized similar tests or, if they existed, whether they were at all reasonable, was a longstanding problem resolved by Linnik and colleagues (1966). This work, in the course of substantial new mathematical developments, showed that no solution existed unless the degrees of freedom were of opposite parity, in which case the solution had critical functions with infinitely many discontinuities. Fisher wrote, in his 1950 author's note on [CP 162], referring to an announced but unpublished proof of non-existence, and responding to criticisms that the Behrens solution did not give a similar test:

> Later S. Wilks has stated that he has proved that no test can exist in this problem, satisfying the conditions laid down by Neyman and Pearson. This, one might have thought, would have settled the matter. It is obviously not an objection to a test of significance that it does not satisfy conditions which cannot possibly be satisfied.

Fisher's reaction to the Linnik results can only be imagined.

Exact confidence solutions are possible by randomization, essentially by pairing observations randomly and using the paired t-test. Fisher argues strongly against this post-experimental randomization (e.g., [SMSI], pp. 98-99). Despite interesting issues of principle and of technique, the topic is tangential both to practice and to the fiducial argument.

The great practical confidence success came in the form of approximations where practical procedures without obvious weaknesses were constructed that came remarkably close to similarity. Welch (1937) provided an approximation based on "Student's" t with degrees of freedom chosen depending on the fraction c.[1] The solution has been widely adopted in textbooks and by practicing statisticians. Later asymptotic refinements by Welch (1947) and Aspin (1948) came closer still to similarity. These will be discussed after more technical development, for their contrast with the Behrens solution is important.

A less clear-cut example will be used as a vehicle to illustrate various procedures, arguments and behaviors. The data are excerpted from a larger randomized clinical trial of several analgesic drugs (Meier, Free and Jackson, 1958). For two drugs administered as the first post-operative drug, the number of whole hours of relief from pain was recorded. The ordered data and summary analysis are set out in Table 2.

Drug 1, used on 7 patients, gave a mean relief of 6.0 hours with a standard error 1.36, whereas drug 2, used on 13 patients, gave a mean relief of 2.23 hours, with standard error 0.60. Because the larger sample variance goes with the smaller sample size, the pooled standard error (1.28) for the difference of means is distinctly smaller than the unpooled standard error $\sqrt{v_1+v_2} = 1.49$ hours.

---

[1] In a brief note, Smith (1936) had earlier proposed the equivalent of the Welch test as an approximation in lieu of tables for Behrens' test. Smith's note was written while visiting the Galton Laboratory and acknowledges Fisher's assistance.

Table 2: Hours of Post-Operative Pain Relief

Drug 1: 2, 4, 4, 5, 6, 8, 13

Drug 2: 0, 0, 0, 1, 1, 2, 2, 2, 3, 3, 3, 4, 8

	Drug 1	Drug 2	Difference
Means	6.00	2.23	3.77
Variances (degrees of freedom)	13.0 (6)	4.69 (12)	--
Standard Error of Mean	1.36	0.60	1.49
$v_i = (SE)^2$	1.86	0.36	2.12

Pooled Standard Error = 1.28

Fraction $c_1 = 1.86/2.12 = 0.84$

What are appropriate 95% inferential limits on the estimated increase of 3.77 hours of drug 1 over drug 2? Figure 2 shows the critical multiples for leading procedures, always in the form of a multiple of the <u>unpooled</u> standard error, here 1.49. The critical multiple depends on the relative sizes of the observed variances, most usefully in the form of the fraction c of the total variance associated with the first mean, here $c = 1.82/2.12 = 0.84$. Critical multiples, in principle readable as ordinates for $c = 0.84$, and the resulting half-widths of the 95% intervals are:

	Multiple	Half-Width
Large sample normal	1.96	2.92
Pooled two-sample t	1.81	2.70
Behrens-Fisher	2.39	3.56
Welch approximate t	2.29	3.41
Welch-Aspin asymptotic	2.32	3.46

The horizontal line for t(18) is displayed for reference but corresponds to no advocated procedure. The pooled-t curve differs from t(18) by the ratio of the pooled to the unpooled standard errors. For equal samples, these two standard errors coincide and the biggest difference among the methods disappears.

By all methods, the difference is significant but the limits are 32% and 26% wider, respectively, for the Behrens and the Welch methods than for the two-sample t method. These differences, about equivalent to 70% increases in sample sizes, do matter for practice.

Alternative analyses for these data are available. A square root transformation makes the sample variances nearly equal, and then the pooled-t is appropriate. In this situation the Welch limits are about 5% tighter and the Behrens limits 5% wider than the pooled t limits.

Differences between procedures can be large or small and can go in all directions except one: the Welch confidence solution always gives tighter limits than the Behrens solution. Such a one-way comparison must come from principled differences between the approaches.

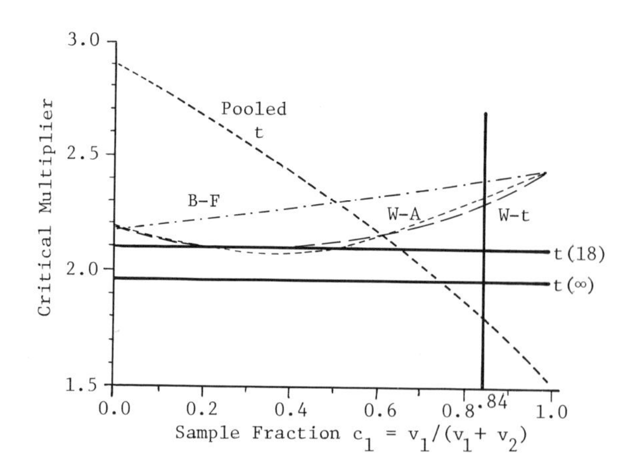

Figure 2. Critical Multiples of Unpooled Standard Error
5% error, degrees of freedom = (6,12)
B-F = Behrens-Fisher; W-A = Welch-Aspin asymptotic;
W-t = Welch approximate t

3. Studentization

By studentization we mean the elimination of a nuisance scale parameter, or more generally, of any nuisance parameters. But what is the inferential procedure?

I start from the premise that we can handle location parameters in the absence of unknown scale parameters, and also inference about a scale parameter from a mean square estimate of it. Consider an inferential statement like

$$|y - \eta| \leq d\sqrt{\Sigma v_i} \tag{1}$$

where we take some multiple, d, of the standard error of the estimated mean, y. How much uncertainty do we attach to this statement about the unknown η? Or what is essentially the same, what level of significance do we attach to an observed y differing from an hypothesized η by d standard errors?

For large samples and for d = 1.96, the answer is approximately 5%. In small samples, the answer depends on how the estimated variance Σv_i compares to the true variance ΣV_i. If the latter were known, the answer is given by a conditional probability calculation, conditional on the sample variance $\{v_i\}$ and the fixed population variances $\{V_i\}$. Thus, define

$$\begin{aligned} U(v,V) &= P\{|y - \eta| \geq d(c)\sqrt{\Sigma v}|v,V\} \\ &= P\{\frac{(y - \eta)^2}{\Sigma V} \geq d^2(c)\,\frac{\Sigma v}{\Sigma V}|v,V\} = \bar{K}\{d^2(c)\,\frac{\Sigma v}{\Sigma V}\} \end{aligned} \tag{2}$$

where $\bar{K}$ is the right-tail cumulative of the chi-square distribution on one degree of freedom.

For "Student's" one-sample problem with only one variance, and with d = 1.96, the uncertainty will be more or less than 5% according as v/V is less or more than 1. We are not ignorant about v/V, for we can make inferences about V from the observed v. If we can justify averaging with respect to a mean square distribution for the ratio v/V, we get an uncertainty slightly over 5%, specifically the two-tail integral of the t distribution beyond d = 1.96. In the confidence approach, the observed sample is imbedded in repetitions for fixed V. Averaging over the mean square distribution of v/V in these repetitions gives the standard t analysis.

The frequency justification is exact within the repetitions for each V. No assumptions about V are required and having extra information about V would not affect the validity of the analysis (though it might make it less than optimal).

The fiducial justification, in contrast, requires the absence of knowledge about V. The averaging is done with respect to a reference distribution of v/V for the observed v. The distribution of v/V is still the same mean square distribution, but with V varying according to the fiducial distribution on the scale parameter. For inference about a location parameter with only one scale parameter, the results of the confidence and fiducial averaging are numerically identical, and differences in the reasoning and the assertions are blurred. For the two-sample problem the differences are sharper.

One difference can be seen already in the one variance problem. Consider the role of information about V, for example, in the form of knowledge that $V > V_0$, as might arise were one component of the error variance known. While that knowledge has no effect on the validity of the confidence construction and interpretation, it does invalidate (or require modification of) the standard fiducial argument. And it should, for it can have a substantial impact on the assessment of uncertainty of the basic assertion (1). Often it will not, of course, if the observed v is much greater than V_0, but it definitely would were the observed v, say, somewhat less than V_0. For then our uncertainty in the statement (1) with d = 1.96 is surely considerably greater than 5%. No averaging can bring that result anywhere near the tail integral of "Student's" t. For me, this captures an essential difference between the inferential approaches. Fiducial and Bayesian inferences are concerned with assessing how much "confidence" can be placed in assertions based on the experiment completed yesterday. Confidence inference changes the problem, guaranteeing 95% correctness for the assertions whose form is chosen today and that will be made about tomorrow's experiments.

Let us turn to the problem of dealing with an estimate $y = \hat{\eta}$ whose variance is the sum of two or more variances. Specializing to exactly two gives only slight simplification that is best made at the end of the argument. Can we choose a multiple d(c) depending on the fractions among the observed variances so that the uncertainty in the assertion (1) will be, say, 5% when we properly account for the unknown population variances?

A fiducial answer is relatively natural and straightforward. Conditional on the population and the observed variances, and for any trial choice d(c), the uncertainty depends on the several ratios v_i/V_i. We must choose d(c) so that the conditional uncertainty averaged over reference distributions on the unknown ratios v_i/V_i will be the desired 5%. The reference distributions depend on what is taken as given. If apart from the experiment we have no knowledge about the individual V_i nor about their interrelationships, "the value v_1 is of no relevance for the estimation of V_2, nor is v_2 of relevance for the estimation of V_1" (Fisher [CP 162, p. 176]). Then the separate fiducial distributions on each population variance V_1 have the form of mean square distributions on the ratios v_i/V_i. We use these independent fiducial distributions as reference distributions to form the average uncertainty. Technically, for each set of $\{v_i\}$, we must choose d(c) to satisfy

$$E_{V|v}\ \{U(v,V)\} = .05,$$

with U as defined in (2). There is a unique choice resulting from explicit, if difficult, numerical computation. The result is identical to that obtained by Behrens' original route summarized in the preceding section. This second route was given in Fisher's second paper on the problem [CP 162] and for me it displays more clearly the fiducial process in studentization.

For the confidence approach, we seek a function d so that, in averages over repeated samples for each fixed V,

$$E_{v|V}\ \{U(v,V)\} \equiv .05.$$

As is typical of "sampling-theory" computations, there is no constructive basis for the choice of function d. As discussed earlier, Linnik (1966) showed that no continuous function d(c) satisfies the conditions. Before these discouraging mathematical results were known, an asymptotic answer was sought in the form of an approximation d_r of d in (2) for which the resulting U_r satisfies

$$E_{v|V}\{U_r(v,V)\} = .05 + 0(1/f^{r+1}).$$

In retrospect, the results were spectacularly successful.

To carry out the first-order asymptotic construction for both confidence and fiducial averaging is relatively easy and leads to instructive comparisons among the leading contending procedures and approximations. Higher-order terms require much more work and are much less instructive. The detailed construction is not essential to an understanding of the discussion and I omit it. It leads to the choice of desired critical functions, through terms of first order in the degrees of freedom, as

$$d_1(c) = z\left(1 + \frac{1+z^2}{4}\,\Sigma\,\frac{c_i^2}{f_i} + \left\{\Sigma\,\frac{c_i(1-c_i)}{f_i}\right\}\right) + 0(f_i^{-2}),$$

where the term in braces is present only for the fiducial solution. In this expression, z denotes the corresponding two-tailed deviate for the Gaussian distribution. For the confidence result, the assertion is only that using this choice of d_1 gives a confidence level that differs from the desired level by an amount uniformly bounded by a term of order f^{-2}; there is no assertion that the critical function approximates to any similar critical function. For the Behrens solution, the approximation is asymptotic to the exact Behrens critical value. Fisher [CP 181] carried the asymptotics for the Behrens solution to three more terms and Aspin (1948) did similarly for the Welch solution, both impressive demonstrations of symbolic algebraic manipulation.

Further simplifications result from noting that "Student's" t on f degrees of freedom has the asymptotic form

$$t_f = z\{1 + \frac{1+z^2}{4f} + 0(f^{-2})\}.$$

If we define an equivalent degrees of freedom f_W by

$$\frac{1}{f_W} = \Sigma\,\frac{c_i^2}{f_i},$$

then the confidence and fiducial solutions can be written, to the same order of approximation, as

Confidence: t_{f_W}

Fiducial: $t_{f_W}\left(1+\Sigma\frac{c_i(1-c_i)}{f_i}\right)$.

The confidence form is exactly that proposed by Welch (1937) and derived in a different manner.

This Welch approximate t solution appears in numerous textbooks, and is often recommended for general use in two-sample comparisons whether variance ratios are seriously unequal or not. Its great advantages over the Welch-Aspin asymptotic constructions are that it requires no special tables beyond the standard t-tables, is well-defined and well-behaved for all choices of degrees of freedom, and has the right limiting behavior as all weight shifts to one variance.

The first-order approximation to the Behrens critical values has the simple property that it is always greater than the Welch solution. Hence it is always a more conservative procedure. The first-order approximation for the Behrens critical values is not very good numerically. Cochran developed another approximation in 1944 for inclusion in Snedecor's Statistical Methods (see Cochran, 1964). Noticing the near linearity in c of the 5% critical values (see Figure 2), Cochran proposed linear interpolation between the limiting t values. In terms of the first order expansions, the approximation can be motivated by noting that for z^2 near 3 (near the 9% level), the effect of the extra term is just to cancel the $\Sigma c^2/f$ term and replace it with a $\Sigma(c/f)$ term instead. To that degree of approximation, a change not of order f^{-2} but of order $(z^2-3)/f$, the critical values take the form

$$d_C = \Sigma\, c_i t_{f_i}.$$

In that form, the Behrens-Fisher test has been recommended for use when variances are known to differ in Snedecor's Statistical Methods (1946; 4th and later editions). Cochran (1964) studied the quality of the approximation, finding it surprisingly good from about 10% down to 1% levels for almost all degrees of freedom. It shares the same practical advantages as the Welch approximate t solution for the confidence problem.

4. Averaged and Conditional Error Levels

For displaying the logic of the fiducial argument and its difference from the confidence argument, averaging over the two independent variances was convenient. For more detailed examination, we can go one more step before the fiducial and confidence arguments diverge analytically. For, were the population variance ratio known to be ρ, then the pooled two-sample solution based on that known ratio would give exact and coinciding fiducial and confidence solutions. Further, the error rates would hold conditionally on the sample and population ratios, r and ρ, or equivalently the fractions c and γ. Technically,

$$M = \left(\frac{f_1 v_1}{\rho} + f_2 v_2\right) / \{V_2(f_1 + f_2)\}$$

is distributed as a mean square on $f_1 + f_2$ degrees of freedom independently of c and γ. Then

$$|y - \eta| \geq d(c)\sqrt{\Sigma v_i}$$

may be rewritten as

$$\frac{|y - \eta|}{\sqrt{M}} \geq d(c)Q(c,\gamma),$$

with Q resulting from the elementary algebra as

$$Q(c,\gamma) = \frac{(1 - c)\gamma f_2 + c(1 - \gamma)f_1}{\gamma(1 - \gamma)(f_1 + f_2)} .$$

If $\bar{H}$ denotes the two-tailed reverse cumulative of "Student's" t on $f_1 + f_2$ degrees of freedom, then

$$P\{|y - \eta| \geq d(c)\sqrt{\Sigma v}|c,\gamma\} = \bar{H}\{d(c)Q(c,\gamma)\}. \tag{3}$$

For any choice of d(c), the confidence error level is

$$\beta(\gamma) = E_{c|\gamma}[\bar{H}\{d(c)Q(c,\gamma)\}],$$

and the fiducial error level is

$$\zeta(c) = E_{\gamma|c}[\bar{H}\{d(c)Q(c,\gamma)\}].$$

The confidence average is taken over the F-distribution of r/ρ for fixed ρ (or over the equivalent distribution in terms of c and γ). The fiducial average is over the same (fiducial) distribution of r/ρ for fixed r. Fisher had naturally formulated his tables and procedures for analysis of variance on the logarithmic scale, and he would refer to averaging over the z distribution of $\frac{1}{2} \ln(r/\rho)$. Snedecor introduced F to protect American agricultural research workers from natural logarithms -- a good step for agriculture but bad for the training of statisticians. In whatever terms, a numerical quadrature is required, for any choice of critical function, to evaluate either the confidence or fiducial levels.

For the 6 and 12 degrees of freedom corresponding to our example, and for the five procedures whose critical functions were displayed in Figure 2, the confidence and fiducial levels have been evaluated and are displayed in Figures 3 and 4. The procedures are the exact Behrens-Fisher solution (B-F), the Welch approximate t (W-T), the Welch-Aspin third-order asymptotic solution (W-A-3), the pooled-t solution assuming equal underlying variances, and the result of using the t(18) critical multiple with the large-sample standard error.

One pair of graphs does not cover the three parameter family of possibilities by varying degrees of freedom and nominal probability levels, but they are typical of all that I have examined. Look first at the confidence levels in Figure 3. Striking features are the nearness to similarity of both Welch solutions, the conservatism of the Behrens-Fisher solution, and the serious deviations for the pooled-t procedure.

Figure 3 illustrates the spectacular success of the Welch-Aspin asymptotic construction. On a finer scale, the maximum level is .0507 for the third order solution, and .0503 for the fourth order solutions. More systematic explorations by its developers and others show comparable results. Despite non-existence of smooth similar tests, the asymptotic construction has yielded a sequence of procedures whose confidence properties are analytically of the desired order and numerically quite close. Terms of order 3 and 4 do begin to become irregular as degrees of freedom go below 6, but much of this is due to the behavior of the asymptotic approximation to t.

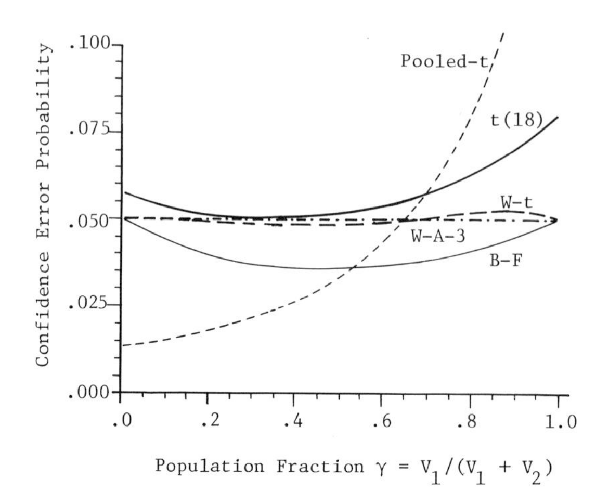

Figure 3. Confidence Error Level Comparisons of Five Procedures
5% nominal level, degrees of freedom = (6,12)

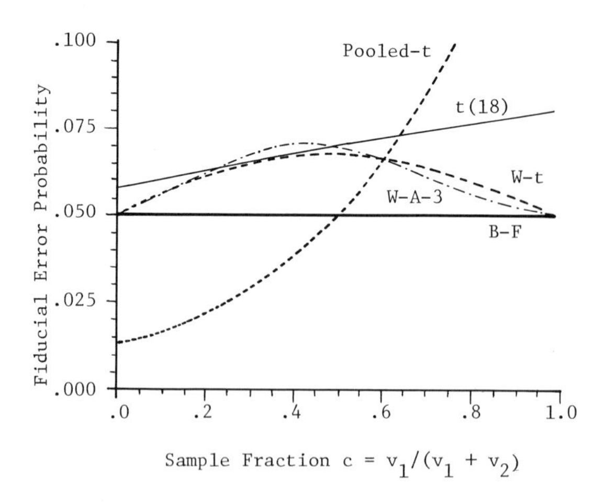

Figure 4. Fiducial Error Level Comparison of Five Procedures
5% nominal level, degrees of freedom = (6,12)

Considered as a confidence solution, the Behrens solution is everywhere conservative, with equality only when the population fraction is 0 or 1. Plotted against the population fraction, the confidence level is not even tangent to the nominal level at the extremes, so that the conservatism is not even close. No analytic proof of the conservatism has been given. Robinson (1976) describes the most systematic numeric exploration, all supporting conservatism. For the history of inference, more needs to be said. In the interchange between Fisher and Bartlett (Fisher [CP 151] and Bartlett, 1936) the conservatism and the fact that the fiducial solution was not a similar confidence solution were shown analytically for each variance with one degree of freedom. The comparison of the first-order asymptotics, noted explicitly by Welch (1947), shows the conservatism for large enough degrees of freedom. In his 1950 Author's note to [CP 162], Fisher asserted:

> It was obvious from the first, and particularly emphasised by the present writer, that Behrens' test rejects a smaller proportion of such repeated samples than the proportion specified by the level of significance, for the sufficient reason that the variance ratio of the populations sampled was unknown.

Our state of understanding is still unsatisfactory. The numerical demonstrations support the truth of Fisher's unsupported assertion, but a proof of why the conservatism must hold is missing.

To Fisher, the unconditional confidence levels were misleading, because of lumping the sample at hand with its observed variance fraction c in with samples with more or less informative fractions. (Which fractions are least informative is not intuitively clear and formed part of the 1936 Bartlett-Fisher interchange.) Averaging instead over the fiducial reference distribution of the population fraction given the observed sample fraction c leads then to the comparisons in Figure 4. The Behrens solution has, by construction, exact fiducial level 5% for all c. The two Welch solutions are anti-conservative by the fiducial standard.

Can the contrasts be made in a way that is more compelling? In 1956, Fisher [CP 264] displayed a conditional behavior of the Welch-Aspin asymptotic solution that he argued should discredit that solution. The conditional computation is just that underlying our derivation of the fiducial solution.

To defend the confidence construction requires that one refuse to grant the necessity and importance of conditioning on the sample variance ratio. Fisher's 1956 message was, in effect: even if you would not accept the Behrens-Fisher averaging, surely you must see something unsatisfactory with the conditional behavior of the Welch-Aspin solution. Many will not or do not see.

Where do we stand on the Behrens-Fisher problem in principle and in practice? The issues of principle remain unresolved and a stumbling block to the foundations of inference.

The effects on statistical practice are also unresolved, but are much less severe, though choices do matter, especially for unequal sample sizes. Even here, the big difference is between using a pooled-t and either a Behrens or Welch procedure. Some of Fisher's strongest defenses of the pooled-t apply solely for testing null differences, and not for setting limits on real differences. But again, his stress on the gains in simplicity and clarity from pooled variances is important.

The Behrens-Fisher solution is a principled solution to the pure problem, conservative, but not clearly flawed. With the Cochran approximation, it is usable in practice without special tables.

The Welch-Aspin asymptotic confidence solution suffers from the need to have special tables, of not being available for very small samples, and for having undesirable conditional behavior. The Welch approximate t solution does not suffer from the first two problems, and avoids the worst conditional misbehavior because its approximate degrees of freedom never exceed the sum of the two components. For equal samples, it is effectively a compromise solution, behaving nearly as pooled-t provided the sample ratio is not too far from one, and providing robustness to major deviations. For unequal samples, it is rather like a compromise solution centered on the "wrong" ratio of underlying variances, roughly n_1^2/n_2^2.

5. The Fieller-Creasy Problem

A method for setting limits on the ratio of two normally estimated means was given by Fieller (1940) in the course of a paper on the standardization of insulin. First included in the 1946 edition of Statistical Methods for Research Workers [SMRW, 1946, Section 26.2], the procedure has become a standard statistical method, especially important in the area of bioassay.

In a symposium of the Royal Statistical Society on interval estimation, Fieller (1954) reviewed the development of the method and some extensions, and Creasy (1954) presented an alternative fiducial solution for the ratio. The ensuing discussion (Irwin et al., 1954) was spirited. The "Fieller-Creasy" problem has become a classical problem of principle in statistical inference, with many more facets than the Behrens-Fisher problem, and its study continues with new life. My selective discussion may stress some atypical aspects.

Fieller (1954) noted that Bliss (1935) had given the solution in a particular case but that Bliss and others had failed to recognize the general applicability. Similarly, Johnson and Neyman (1936) had also given the exact analysis in the form of a test for the values of matching variables for which the ordinates of two regressions are equal. Their general analysis, fully valid for any number of matching variables, was specifically explored only for two variables and the richer value of the one-variable applications was never noted.

In the usual form of the abstract problem, x_1 and x_2 are observed, independently and normally distributed with means μ_1 and μ_2, and our interest is in the ratio $\lambda = \mu_2/\mu_1$. Taking both variances as known and equal to one leaves intact the essential problems. Any hypothesis specifying a value λ of the ratio can be written as a linear hypothesis $\mu_2 - \lambda\mu_1 = 0$ for which

$$Q_\lambda = \frac{(x_2 - \lambda x_1)^2}{1 + \lambda^2}$$

is the standard test statistic, distributed as χ_1^2 for all μ_1, μ_2 with the given ratio.

Inverting the test gives limits for the value of the ratio consistent with the data. The behavior of Q_λ as a function of λ is illustrated in Figure 5. For

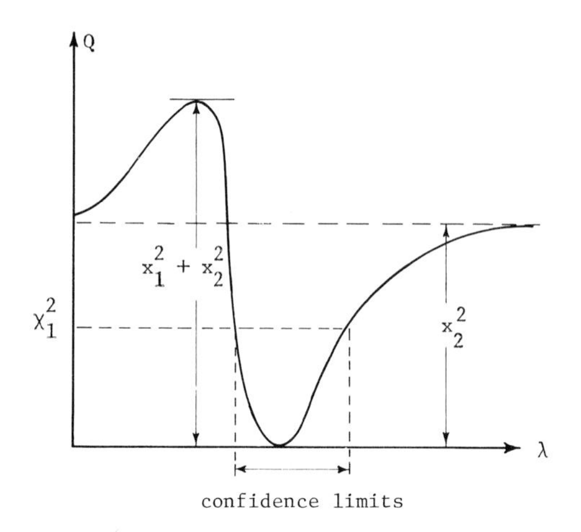

Figure 5. Test Statistic Q as Function of Hypothesized Ratio λ

critical values of χ_1^2 less than x_1^2, the inversion gives an interval for λ. For values of χ_1^2 between x_1^2 and $x_1^2 + x_2^2$, the inversion gives the complement of an interval for λ. For values larger than $x_1^2 + x_2^2$, no value of λ is rejected. These phenomena have long been subjected to much discussion, to which I shall return later.

The resulting "limits" on λ comprise the Fieller solution. They yield a nested set of confidence statements as the confidence level increases. For any fixed level, the statements have exact frequency of correctness in repeated samples from population means in any ratio. The limits were given as fiducial limits by Fieller and they are called fiducial limits by Fisher, but in his 1954 discussion [CP 257], he stresses that he did not describe the result as a fiducial distribution for the ratio. The Fieller limits are surely not Bayesian limits for any continuous prior because for each sample, the limits take in the whole space for some positive error level, inconsistent with a conditional statement of posterior probability less than one. That this same conditional phenomenon does not cause problems for fiducial inference should be unsettling.

Creasy (1954) uses the bivariate normal fiducial distribution on (μ_1,μ_2) induced by (x_1,x_2), and proceeding analogously to finding the fiducial distributions for $\mu_1 - \mu_2$, constructs the marginal distribution of the ratio μ_2/μ_1. Her construction is evidently the Bayesian analysis for the invariant prior $d\mu_1 d\mu_2$. Her solution differs from Fieller's in a uniform and interesting way. For a pair of limits on the ratio such as shown in Figure 6, the fiducial probability given by Creasy is just the probability under the bivariate normal distribution in the areas A and C. In contrast, the fiducial probability under the Fieller solution amounts to the probability in A less that in C.

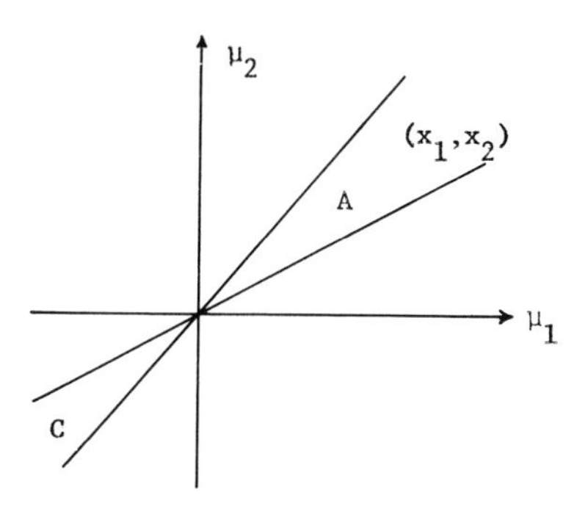

Figure 6. Regions in Limits on Ratio of Means

Creasy's limits, being always tighter than Fieller's, cannot be confidence limits and will everywhere have a coverage smaller than the nominal level. Whether Creasy's solution is a fiducial solution depends on who is judge. The fiducialists among the discussants all defended Fieller and attacked Creasy. Fisher's contribution is described in Joan Fisher Box's biography (1978, p. 459):

> ... paper ... which Fisher denounced. This is a case in which Fisher was clearly unfair; he gave no criterion for his rejection of one of the sets of pivotals Creasy used.

The worst aspect of Fisher's discussion is the inconsistency between its technical content, which comes close to repetition of Creasy's analysis, and its evaluative critical attack on her.

But Creasy's solution and Fisher's discussion are both diversions. Fisher reminded Creasy and us that the analysis must depend on the problem and what can be assumed as given. Most of the symposium discussion was, I believe, wasted in disputation over premature abstractions of the ratio problem.

In his contribution to the discussion, Barnard tied his justification for using the area A - C rather than A + C to the special role of the origin in determining whether the ratio is defined. That, too, cannot be resolved by abstract argument, but does lie at the heart of the problem, and of several recent developments.

Scheffé (1970) proposes a new confidence solution and James, Wilkinson and Venables (1974) (hereafter called J-W-V) propose a new fiducial solution, each more conservative than Fieller's and _a fortiori_, than Creasy's. Each of these start from the idea that the ratio is meaningless if the true mean is (0,0) and that no informative inference should then be made about the _ratio_. Excluding no value of the ratio at, say, the 5% level is not a weakness but an essential desideratum. Within the spirit of Scheffé's multiple comparisons development, one should not get less than the full line for the ratio until $x_1^2 + x_2^2$ exceeds the 5% value for chi-square on two degrees of freedom rather than for the one-degree of freedom occurring in Fieller's method.

The J-W-V solution is an instructive example of serious fiducial construction, not dissimilar to the analysis for the Behrens-Fisher problem, but satisfying stronger conditions set down by Wilkinson (1978). While the J-W-V and Scheffé solutions have major technical and philosophical differences, their numerical results are very similar. Under some conditions, both differ greatly from Fieller's results.

The importance of the origin and of ratios suggests that a reformulation of the problem in polar coordinates would be preferable for clarifying the analysis. (The J-W-V development is wholly in polar coordinates.) Let $r = \sqrt{x_1^2 + x_2^2}$, $\rho = \sqrt{\mu_1^2 + \mu_2^2}$, $u = \tan^{-1}(x_2/x_1)$, and $\theta = \tan^{-1}(\mu_2/\mu_1)$. Interest in ratios corresponds to interests in the angle θ (modulo 180^0), with the radius ρ a nuisance parameter. Fieller's solution takes the simple form

$$r^2\sin^2(\theta-u) \leq \chi_1^2.$$

Unless r^2 exceeds χ_1^2, no angle is ruled out; otherwise, the method gives symmetric limits about the observed angle. The asymmetry of the Fieller and other solutions is an artifact of bad parameterization. Similarly, the limits excluding an interval correspond to limits on the angle including the vertical direction.

The Creasy, Scheffé, and J-W-V procedures all yield limits for the angle θ that are symmetrical about the observed angle. All require numerical integrations and special tables or approximations. Figure 7 displays the half-width of the 95% limits for the (unoriented) angle for the four procedures, each as a function of the observed radius $r = \sqrt{x_1^2 + x_2^2}$. The differences among the procedures is concentrated in samples for which the origin cannot be overwhelmingly rejected.

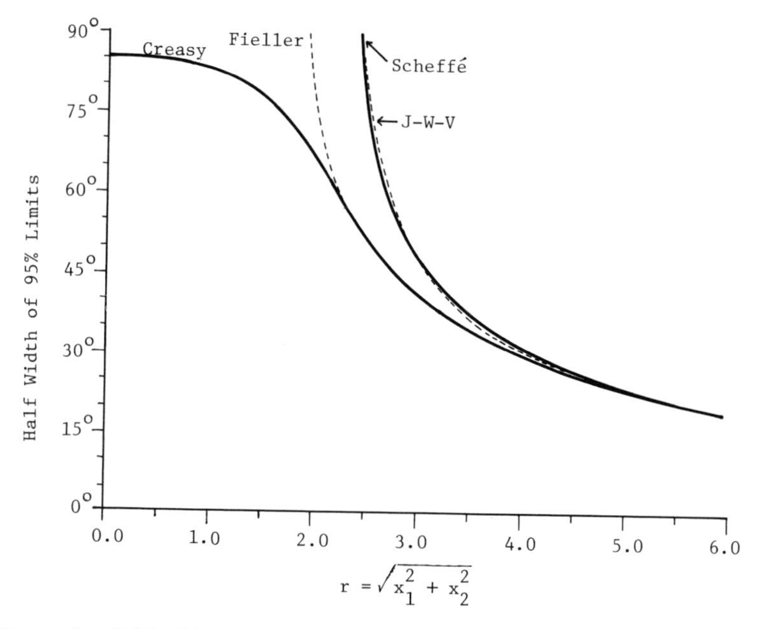

Figure 7. Half-width of 95% Limits on Unoriented Angle for Four Procedures

Scheffé specifies that when the limits encompass all angles, then the form of the inference reverts to an elliptical (here circular) region on the bivariate population mean, a region that will enclose the origin. For J-W-V, the phenomenon is represented by a fiducial probability associated with the origin. By whichever route, the two new solutions move away from the inadequate formulation of treating the inference solely in terms of the angle (or ratio). Only the Bayesian position is unrepresented by a broadening of viewpoint.

In problems where ratios are properly thought of in terms of unoriented directions from a special origin, the J-W-V and Scheffé solutions appear satisfactory, and their arguments compelling. The J-W-V development applies as easily to inference for oriented directions (rays). But the most common problems for which Fieller's method have been used are not of this form. Consider two. In indirect assay, parallel linear dose-response regressions are fitted for the standard and test drugs. The dose z_U of the test drug with the same effect as dose z_S of the standard is determined by

$$a_S + bz_S = a_U + bz_U$$

$$z_U - z_S = \frac{a_S - a_U}{b} = \frac{x_2}{x_1} .$$

The interest is in the ratio. The zero of the denominator matters immensely; the zero of the numerator is ordinarily of no import. The entire vertical axis is singular and the origin is not special.

Before the indirect assay would be used, preliminary tests would be made for parallelism, for linearity and for significantly positive slope. Without satisfactory answers to all three, the estimates given above would not be used. The first two preliminary tests tend to drop from sight once the model is adopted. The potential failure of the third remains as part of the model, and Fieller's limits on the ratio fail to be finite inclusive intervals exactly when the slope (the denominator) is not significantly different from zero. Scheffé and J-W-V's argument that $x_1^2 + x_2^2 = \chi_2^2$ should be the breakdown point loses its force, but not their idea that no informative inference about the ratio should be made if it is truly indeterminate. The breakdown point for inference about the ratio should be $x_1^2 = \chi_1^2$, and the exclusive interval should not be used. Instead, perhaps, the inference should be made in the form of limits (including 0) on the slope, or in the form of fiducial probability associated with a zero slope. Without expanding the exclusive limits to the full line, one does not have the confidence coverage that for zero slope, failure to cover every value of the ratio is an "error". That the resulting limits have a discontinuity as the error level diminishes suggests that this adjustment of the Fieller solution is not the final solution.

The preceding considerations apply with equal strength to a second problem: setting limits on the location of the minimum of a quadratic regression. The location is given by a ratio of regression coefficients. Again, the zero of the denominator, the quadratic coefficient, is crucial; the zero of the numerator, the linear coefficient, is irrelevant. Further, if the denominator is negative, the ratio gives a maximum, and the fitted curve has no finite minimum.

The mathematical structure is preserved by reformulating the question to setting limits on the stationary point, but the statistical problem is usually destroyed. What is needed is really some adequate formulation of inference to give limits for a construct if it exists and to do something else sensibly if it does not exist. Some combination of preliminary tests and Fieller's solution provide generally reasonable procedures. Fieller (1954) considered some aspects of the questions raised by the extension of the minimum problem to cubic regressions, but intriguing issues remain.

7. Conclusion

The Behrens-Fisher and Fieller-Creasy problems have not been settled, and challenges to principled formulations and wise practice remain, not least for the Bayesians. Fiducial ideas, especially as they relate to these problems, have not lost their suggestive strength. In all of our consideration and extension of statistical methodology and principles, we can profit from keeping close to the spirit of Fisher's insistence on working from specific problems.

References

Aspin, A.A. (1948). "An Examination and Further Development of a Formula Arising in the Problem of Comparing Two Mean Values," Biometrika, 35, 88-96.

Bartlett, M.S. (1936). "The Information Available in Small Samples," Proceedings of the Cambridge Philosophical Society, 32, 560-566.

Behrens, W.V. (1929). "Ein Betrag zur Fehlerberechnung bei weniger Beobachtungen," Landwirtschaftliche Jahrbücher, 68, 807-837.

Bliss, C.I. (1935). "The Calculation of the Dosage-Mortality Curve," Annals of Applied Biology, 22, 134-167 and 307-337.

Box, Joan Fisher (1978). R.A. Fisher: The Life of a Scientist. New York: John Wiley and Sons.

Cochran, W.G. (1964). "Approximate Significance Levels of the Behrens-Fisher Test," Biometrics, 20, 191-195.

Creasy, M.A. (1954). "Limits for the Ratios of Means," Journal of the Royal Statistical Society, Series B, 16, 186-194.

Fieller, E.C. (1940). "The Biological Standardization of Insulin," Supplement to the Journal of the Royal Statistical Society, 7, 1-64.

Fieller, E.C. (1954). "Some Problems in Interval Estimation," Journal of the Royal Statistical Society, Series B, 16, 175-185.

Irwin, J.O., et al. (1954). "Discussion on the Symposium on Interval Estimation," Journal of the Royal Statistical Society, Series B, 16, 204-222.

James, A.T., Wilkinson, G.N., and W.N. Venables (1974). "Interval Estimation for a Ratio of Means," Sankhyā, A, 36, 177-183.

Jeffreys, H. (1940). "Note on the Behrens-Fisher Formula," Annals of Eugenics, 10, 48-51.

Johnson, Palmer O. and J. Neyman (1936). "Tests of Certain Linear Hypotheses and Their Application to Some Educational Problems," Statistical Research Memoirs, 1, 57-93.

Linnik, J.V. (1968). Statistical Problems with Nuisance Parameters. Translated from 1966 Russian Edition. Providence: American Mathematical Society.

Meier, P., Free, S.M. and G.L. Jackson (1958). "Reconsideration of Methodology in Studies of Pain Relief," Biometrics, 14, 330-342.

Rayleigh, Lord (1893). "On the Densities of the Principal Gases," Proceedings of the Royal Society of London, 53, 134-149.

Rayleigh, Lord (1894). "On an Anomaly Encountered in Determinations of the Density of Nitrogen Gas," Proceedings of the Royal Society of London, 55, 340-344.

Robinson, G.K. (1976). "Properties of Student's t and of the Behrens-Fisher Solution to the Two Means Problem," Annals of Statistics, 4, 963-971.

Scheffé, H. (1970). "Multiple Testing Versus Multiple Estimation," Annals of Mathematical Statistics, 41, 1-29.

Smith, H.F. (1936). "The Problem of Comparing the Results of Two Experiments with Unequal Errors," Council of Scientific and Industrial Research (Australia), Journal, 9, 211-212.

Snedecor, G.W. (1946). Statistical Methods, 4th Edition. Ames: Iowa State College Press.

Welch, B.L. (1937). "The Significance of the Difference Between Two Means when the Population Variances are Unequal," Biometrika, 29, 350-362.

Welch, B.L. (1947). "The Generalization of "Student's" Problem when Several Different Population Variances are Involved," Biometrika, 34, 28-35.

Wilkinson, G.N. (1977). "On Resolving the Controversy in Statistical Inference (With Discussion)," Journal of the Royal Statistical Society, Series B, 39, 119-171.

This research was supported by U.S. Department of Energy Contract No. EY 76-S-02-2751 to the University of Chicago.

FISHER, JEFFREYS, AND THE NATURE OF PROBABILITY

David A. Lane

1. Introduction

Bayesian inference was Fisher's intellectual bête noire. His work is filled with sharp attacks on "inverse probability",[1] and for many years he sought to develop a logic of induction alternative to the Bayesian scheme. In this connection he developed some of his central ideas, including the likelihood-based theory of estimation and the theory of fiducial probability. This lecture is about Fisher's critique of Bayesian inference. It centers on an acerbic and amusing exchange between Fisher and Jeffreys which took place in the 1930s and revealed which parts of the Bayesian theory Fisher rejected, and which parts he simply failed to comprehend. First, I shall briefly summarize two papers on inverse probability which antedate this exchange, and at the end of the lecture I shall mention a few later changes in point of view which brought Fisher more into line with the ideas argued by Jeffreys in their exchange -- but which Fisher had rejected out of hand at the time.

2. Fisher's Early Criticism of Bayesian Inference

> Inverse probability has, I believe, survived so long in spite of its unsatisfactory basis, because its critics have until recent times put forward nothing to replace it as a rational theory of learning by experience. [CP 84, p. 531].

In the 1920s and early 1930s, Fisher advanced a new framework for a "rational theory of learning by experience", based upon likelihood and fiducial probability. To clear the field for his ideas, he launched an attack in the papers "Inverse

---

[1] See, just for example, in addition to the works discussed in this talk, [CP 137] and [CP 273], as well as the introduction to Design of Experiments [DOE] and the second chapter of Statistical Methods and Scientific Inference [SMSI]. Bayes himself is the second most-cited statistical author in the Collected Papers.

Probability" [CP 84] and "Inverse Probability and the Use of Likelihood" [CP 95] designed to expose the "unsatisfactory basis" of the older rival theory, inverse probability.

He identified three major fallacies underlying inverse probability. First, the theory of inverse probability required every inference to be couched in the language of probability:

> ... we learn by experience, science has its inductive processes, so that it is naturally thought that such inductions, being uncertain, must be expressible in terms of probability. In fact, the argument runs somewhat as follows: a number of useful but uncertain judgments can be expressed with exactitude in terms of probability; our judgments respecting causes or hypotheses are uncertain, therefore our rational attitude towards them is expressible in terms of probability. [CP 84, p. 28].

But this argument was faulty according to Fisher. In particular, inferences from a sample to a population should often be expressed not in terms of probability, but in terms of likelihood, which provides "as satisfactory a measure of 'degree of rational belief' as a probability could do" [CP 95, p. 259]. Of course, *which* inferential situations call for probabilistic expression must depend on how one defines "probability". Fisher did not explicitly consider this question in [CP 84] or [CP 95], but his ideas on a proper definition of probability -- and the consequent reduction of its domain of applicability -- emerged clearly enough in the exchange with Jeffreys. The Prefatory Note in [CP 42] gives a statement of Fisher's early notion of the "hypothetical infinite population" and its relation to probability.

The remaining two fallacies of inverse probability involve the assignment of an *a priori* distribution to population parameters. When the population parameters $\underset{\sim}{\theta}$ are in fact sampled from a superpopulation with known sampling distribution, Bayes Theorem -- which is, after all, a theorem -- can be applied with perfect validity. The second fallacy of inverse probability involves situations in which the *a priori* knowledge of $\underset{\sim}{\theta}$ is not of this "definite kind" [CP 95, p. 257]: then Fisher argued that any assignment of an a priori distribution to $\underset{\sim}{\theta}$ is *arbitrary* and hence invalid. To the extent that such an assignment is based upon some assessment of "equally possible cases", it "reduces all probability to a subjective judgment" [CP 84, p. 528]. This introduction of an arbitrary and subjective element into probability calculations was clearly intolerable to Fisher.

Asymptotic considerations cannot paper over this fundamental weakness in the Bayesian scheme. The arbitrariness involved in assigning a distribution to $\underset{\sim}{\theta}$ is not mitigated by the mathematical fact that "as the observational material is made more and more ample, uncertainty with respect to our *a priori* premises, makes in our result less and less difference" [CP 95, p. 257]. This fact provides no guarantee that the assignment is harmless "when our observations are finite in number, as is invariably the case in practice" [CP 95, p. 258]. But even if the assignment turned out to be completely irrelevant, this should suggest "that conclusions can be drawn from the data alone, and that, if the questions we ask seem to require knowledge prior to these, it is because ... we have been asking somewhat the wrong questions" [CP 95, p. 258]. So assigning *a priori* distributions in the absence of sampling information about the parameters either affects resulting inferences -- in which case it is misleading; or does not, in which case it is an irrelevant and wrongheaded procedure.

The third fallacy which Fisher detected in the foundations of inverse probability concerned the extension of Bayes Theorem by "Bayes Axiom" [DOE, p. 6], which specified a particular choice of *a priori* distribution:

> The peculiar feature of the inverse argument proper is to say something equivalent to "we do not know the function ψ specifying the superpopulation, but in view of our ignorance of the actual values of $\underset{\sim}{\theta}$, we may take ψ to be constant. [CP 84, p. 531].

This "axiom", Fisher argued, renders Bayesian inference inconsistent, since if $\underset{\sim}{\theta}'$ represents a reparameterization of $\underset{\sim}{\theta}$, $\underset{\sim}{\theta}'$ and $\underset{\sim}{\theta}$ cannot in general both have constant densities, although the "axiom" implies that they both should. But is the axiom a necessary component of the theory of inverse probability? Certainly not, when probabilities are defined in personalistic terms as in Ramsey's treatment (Ramsey, 1926). Nor did Jeffreys invoke "Bayes axiom" in the form cited by Fisher. In fact, the whole Fisher-Jeffreys exchange was precipitated by an attempt on Jeffreys' part to derive an *a priori* distribution, in terms of a particular parameterization, with a particular desirable property -- an attempt which would have no justification had Jeffreys subscribed to the Fisherian form of "Bayes axiom". While the use of a "nonsampling" prior is essential to inverse probability, "Bayes axiom" and the consequent inconsistency in inference is not.

It is unfortunate that in his criticism of inverse probability Fisher did not explicitly refer to the work of any particular statistician. However, in the exchange with Jeffreys, he was forced to confront an intelligent scientist rather than a strawman "inverse probabilist" of his own creation.

3. The Fisher-Jeffreys Exchange

The Fisher-Jeffreys exchange consisted of five papers in the Proceedings of the Royal Society of London. The first paper appeared in 1932, the last two in 1934.

3.1. Jeffreys' 1932 Paper. The exchange began with an argument developed by Jeffreys during a treatment of the measurement error problem when "the normal law of error is supposed to hold". Jeffreys asked: what should be taken for the prior probability distribution on the parameters μ and σ, when these are "supposed completely unknown"? (Jeffreys, 1932, p. 48). Considerations which need not concern us led Jeffreys to a prior of the form $f(\sigma)d\mu d\sigma$. The question he addressed, then, was: How to determine f?

Jeffreys reduced this problem to a simpler one, the answer to which he considered self-evident -- or at any rate, derivable from an easy application of the principle of insufficient reason -- and which was consistent with one and only one function f. He posed the question: Having taken two observations, x_1 and x_2, "what is the probability that the third observation [x_3] will lie between them?" His solution: "The answer is easily seen to be 1/3. For the law says nothing about the order of occurrence of errors of different amounts, and therefore the middle one in magnitude is equally likely to be the first, second, and third made (provided, of course, that we know nothing about the probable range of error already)." Computing the posterior distribution for x_3, Jeffreys found that he must set $f(\sigma) = 1/\sigma$ in order to guarantee that

$$P[\min(X_1,X_2) \leq X_3 \leq \max(X_1,X_2)|X_1 = x_1,X_2 = x_2] = 1/3$$

for all values x_1 and x_2.

If Jeffreys' arguments were valid, two of Fisher's points of attack against Bayesian inference would be undermined. First, Jeffreys had produced a unique prior distribution for a "completely unknown" parameter, without recourse to "arbitrary" or "subjective" procedures. Second, the resulting prior density for σ, in contradistinction to the purported "Bayes axiom", was not constant. So Fisher fired off a retort, designed to discredit Jeffreys' argument and to put forward a correct fiducial analysis of the normal measurement problem in its place.

3.2. Fisher's First Reply [CP 102]. Fisher began by recasting Jeffreys' problem in Fisherian language:

> Jeffreys considers the question of: what distribution a priori should be assumed for the value of σ, regarding it as a variate varying from population to population of the ensemble of populations which might have been sampled?

It is important to realize that this is not the problem which Jeffreys considered. The distinction between Fisher's and Jeffreys' version of the question lay at the heart of the disagreement between the two. For Jeffreys, the prior distribution of σ encoded a particular state of knowledge about a particular measurement procedure. Fisher, on the other hand, seemed to interpret the prior distribution for σ as a statement about the frequency distribution of measurement procedures in the world. Fisher, not Jeffreys, introduced the notion of a "population" of measurement errors, "sampled" according to some probability law from an "ensemble" of populations, each population associated with its standard deviation σ. Unfortunately, Fisher's criticisms dealt entirely with his version of Jeffreys' problem, and so rather missed the point of Jeffreys' argument. Nonetheless, Fisher's paper revealed an important element of his thinking at this stage in his career: he insisted upon interpreting every probability statement as a statement about the frequency of some attribute in some population, not as a measure of degree of belief -- even when the only conceivably relevant population was "hypothetical" and "infinite".

Given his interpretation of the problem, it was easy for Fisher to produce a "counterexample" to Jeffreys' argument. After all, how could one expect to discover anything about the distribution of measurement procedures "by mathematical reasoning only without recourse to observational material?" Suppose you set up "an artifically constructed series of populations having any chosen distribution of σ": then Jeffreys' distribution for σ would be different from the real population distribution of σ's! So Jeffreys must be wrong.

Where, according to Fisher, did Jeffreys' argument go astray? Here are two probability statements:

(a) $P[\min(X_1,X_2) \le X_3 \le \max(X_1,X_2)|\mu,\sigma] = \frac{1}{3}$

(b) $P[\min(X_1,X_2) \le X_3 \le \max(X_1,X_2)|X_1 = x_1,\ X_2 = x_2, \mu,\sigma] = \frac{1}{3}$

Clearly, (a) is true. Fisher claimed that Jeffreys had confused (b) -- which is certainly not true in general -- with (a), and then had asserted the truth of (b). In Fisher's words:

> ... we may note at once that, for any particular population, the probability [that the third observation will lie between the first two] will generally be larger when the first two observations are far apart than when they are near together. This is important since, as will be seen, the fallacy of Jeffreys' argument consists just in assuming that the probability will be 1/3, independently of the distance apart of the first two observations. [CP 102, p. 79].

In fact, exactly what Jeffreys was attempting to describe was a state of knowledge about σ under which, no matter what values x_1 and x_2 are observed, it is impossible to say whether these values are "near together" or "far apart". This state of knowledge implies neither (a) nor (b), but it does suggest the truth of statement (c) for all x_1,x_2:

(c) $P[\min(X_1,X_2) \le X_3 \le \max(X_1,X_2)|X_1 = x_1,\ X_2 = x_2] = \frac{1}{3}$,

and this is what Jeffreys assumed to be true. Fisher's insistence on a frequency interpretation of probability statements allowed no direct interpretation of statement (c). His imputation of statement (b) to Jeffreys initiated his criticism of Jeffreys' argument.

Fisher concluded his paper by deriving the fiducial distribution of σ. This distribution coincided with the posterior distribution calculated from the observations and Jeffreys' prior. Nonetheless, Fisher considered the two distributions "logically entirely distinct", and did not consider why the two distributions agreed. As a parting shot, he attacked Jeffreys' prior for its impropriety:

> It is, moreover, as Jeffreys ... clearly perceives, an impossible distribution *a priori*, since it gives zero probability *a priori* for σ lying between any finite limits, however far apart. In the fiducial form of statement this difficulty does not occur. [CP 102, p. 83].

3.3. *Jeffreys' Rejoinder*. Fisher had translated Jeffreys' argument into a foreign inferential language. Before any further discussion about this particular problem could ensue, the underlying disagreement between the two scientists about the nature of probability and induction had to be made explicit. Jeffreys (1933) attempted this task in his rejoinder to Fisher. The main theme of his essay was that probability is the correct mode for all inductive inference, but not probability-according-to-Fisher: "By 'probability' I mean probability and not frequency, as Fisher seems to think, seeing that he introduces the latter word in restating my argument".

Probability is a "relation between one proposition p and another proposition (or aggregate of propositions) q expressing the *degree* of knowledge concerning p provided by q". Thus probability statements do not merely describe the frequency of attributes in aggregates, but represent "an expression of our state of knowledge of propositions in relation to particular data". Regarding the mathematical properties of probability, if one assumes comparability and transitivity of the qualitative "more probable than" relation, probabilities can be measured by real numbers between 0 and 1, in such a way that they satisfy the fundamental laws of additivity and Bayes Theorem. Jeffreys cited his own work (in particular *Scientific Inference*, 1931) and Ramsey's (1926) "Truth and Probability" in support of this contention.

The mathematical theory for manipulating probability numbers is, however, not sufficient. At some point, one must develop

> ... rules for deciding what numbers are to be put into it. The fundamental rule is the Principle of Nonsufficient Reason, according to which propositions mutually exclusive on the same data must receive equal probabilities if there is nothing to enable us to choose between them.

This principle may be applied in particular problems to yield *a priori* distributions of probability. The application of the principle may be direct, or, as in Jeffreys' previous paper, "we may investigate what distribution is consistent with facts otherwise known about the posterior probability on certain types of data".

Now Fisher had derided the possibility of assigning a priori distributions for parameters, even though he computed fiducial distributions for them based upon observations. To Jeffreys, these positions contradicted one another, since it is a simple consequence of Bayes Theorem that "if we can assign a meaning to a probability on experience, we can certainly assign one without it"; just use the fact that posterior odds equal the product of likelihood ratio and prior odds.

In fact, claimed Jeffreys, it was Fisher's conception of probability in terms of frequencies in hypothetical infinite populations which was impossible to interpret sensibly; see Chapter 7 of Jeffreys (1961) for a later discussion of this criticism. "The simple answer to Fisher's theory is that the hypothetical infinite population does not exist, that if it did its properties would have to be inferred from the finite facts of experience, and not conversely, and that all statements with respect to ratios in it are meaningless." Thus, far from demolishing the Bayesian structure in general, and Jeffreys' application of it in particular, Fisher had simply failed to understand the proper role of probability and had developed no coherent theory to put in its place.

Jeffreys also offered rebuttals to some of Fisher's more specific criticisms. First, he dismissed Fisher's "counterexample" based upon the frequency interpretation of the prior distribution of σ: "Fisher considers that it is an objection to my theory that with different assumptions I should have got a different answer, and thereby misses the entire point of the theory". If definite "sampling information" about σ existed, "there would, of course, be no need for any further discussion of the form of f; if it is known, it is known, and there is no more to be said".

Second, the impropriety of Jeffreys' prior is not a defect, as Fisher implied, but a virtue. Using it, the posterior for σ after one observation is still $d\sigma/\sigma$.

> This is what we should expect, since a single observation can tell us nothing about its own precision. It is only when we have two observations that a definite standard of precision, given by their separation, is introduced, and it is precisely then that the theory for the first time leads to definite results as to the probability [that σ lies within a stated interval].... At all points the theory gives results in accordance with expectation.

Third, Fisher's claim that Jeffreys should properly have considered only the unconditional form, probability statement (a), was incorrect:

> Fisher proceeds to reduce my theory to absurdity by integrating with respect to all values of the observed measures. This procedure involves a fundamental confusion, which pervades the whole of his statistical work, and deprives it of all meaning. The essential distinction in the problem of inference is the distinction between what we know and what we are trying to find out: between the data and the proposition whose probability on the data we are trying to assess. If we have made two observations, ... those are our observations, and there is no more to be said. To integrate with respect to them and average a function of them over the range of integration is an absolutely meaningless process. Yet in Fisher's constructive, as well as his destructive work, this process is carried out again and again.

Finally, Jeffreys rejected Fisher's fiducial argument. Given two spaces Ω_1 and Ω_2 and a conditional distribution on Ω_2 with respect to each point on Ω_1, a probability distribution on $\Omega_1 \times \Omega_2$ is not determined: it does not help to say that the marginal on Ω_1 is unknown. It must be <u>assessed</u>, as Jeffreys had attempted in a particular case in his earlier paper.

3.4. <u>Fisher's Second Reply [CP 109]</u>. Fisher's response to Jeffreys' essay appeared in 1934. In this reply, he attacked Jeffreys' definition of probability, denied Jeffreys' contention that probability sufficed for expressing all inductive inferences, and defended his own conception of the "hypothetical infinite population" and its inferential role.

Jeffreys' "subjective and psychological" definition of probability failed for three reasons. First, if probability refers to a state of knowledge, it is not an attribute of the physical world with an objective value, "in the sense that the weight of an object, and the resistance of a conductor have objective values". This criticism would have told more strongly against Jeffreys had Fisher explained just which entities in the physical world possess objective values of probability, and by what operations these values may be assessed.

Second, Jeffreys' epistemic probability statements cannot be tested against observation: "his rejection also of frequency as an observational measure of probability makes it impossible for any of his deductions to be verified experimentally". This criticism is hard to understand. Certainly Jeffreys' definition of probability is not operational (although other epistemic, non-frequentist theories -- notably the developments of Ramsay and de Finetti -- attempt to define probability for observations given hypotheses, probabilities which led to predictions of frequencies in future experiments). These predictions are as open to experimental verification as any frequentist probability statement can be (for a discussion of the relation between epistemic probability and the prediction of frequencies, see Jaynes (1974)).

Finally, Fisher claimed that the Principle of Nonsufficient Reason "leads to inconsistencies which seem to be ineradicable". But he supported this charge only with an ambiguous example due to Keynes, already considered by Jeffreys.

Certainly a definition of probability which cannot be applied to objects of fundamental scientific interest, which allows for no connection with scientific experimentation and which implies logical inconsistencies, is a bad definition. This was Fisher's indictment of Jeffreys' definition; but that he substantiated it is doubtful.

Can a "single quantity ... provide a measure of 'degree of knowledge' in all cases in which uncertain inference is possible?" Jeffreys believed probability did so, but Fisher presented two situations in which he believed another quantity was required. First, suppose one wishes to measure "the amount of information which the proposition, or set of propositions, q, has to offer respecting the truth or falsehood of p"; in the estimation setting, the appropriate measure would be quantity of information, not probability. Second, again in the estimation setting, all available a priori knowledge may be exhausted in "the specification of the forms of population" to be considered, and nothing may be known which can differentiate between possible parameter values. In this case, likelihood is appropriate for forming inferences and population parameters for observations. Probability, claimed Fisher, is simply not available.

Fisher's defense of his own definition of probability was too sketchy to be useful. He made a distinction between frequencies in "hypothetical infinite populations" -- upon which his definition was based -- and frequencies of experimental events, but he did not say what "hypothetical infinite populations" are, what their relationship to real problems of interest might be, or how they could be studied. In summary, Fisher's answer to Jeffreys -- its constructive as well as its critical aspects -- seems incomplete and unconvincing.

3.5. Jeffreys' Final Rejoinder. Jeffreys' answer appeared in the same 1934 issue of the Proceedings of the Royal Society of London. In it, he attempted to clarify his notion of probability and defend it against Fisher's attack. His most important point came in response to Fisher's claim that Jeffreys' probabilities were not experimentally verifiable. This objection, Jeffreys argued, overlooked the fact that no theory of induction could be free of *a priori* assumptions:

> It is not claimed that either the laws of probability or the assessment of prior probabilities can be proved, either by logic or by experiment. If they could be proved, the belief that inductive inference is meaningless would be disproved, and this seems impossible. All we can say is that we do believe that inference is possible, and therefore take this as an axiom. It is a question whether a self-consistent theory can be constructed with a given set of postulates, but enough seems to have been done now [despite Fisher's claim to the contrary!] to entitle us to answer this question in the affirmative.

Prior probabilities were not intended to lead to verifiable predictions: "A prior probability is not a statement about frequency of occurrence in the world or in any portion of it. Such a statement, if it can be established at all, must involve experience. The function of the prior probability is to state the alternatives to be tested in such a way that experience will be able to decide between them."

Does this mean that prior probabilities were subjective -- whether or not one agreed with Fisher in assigning this word a perjorative connotation? Jeffreys considered this question irrelevant. For the distinction between subjective and objective involved the question of the nature of reality, and Jeffreys contended that the interpretation of reality "involves the whole procedure of inference, which is therefore more fundamental in knowledge than any question of reality, which is not very important for scientific purposes"; see Chapter 8 of Jeffreys (1961) for further discussion. Fisher had asserted that a probability might have "objective value" which Jeffreys' theory could not determine; Jeffreys responded, in essence, that to come to understand how we might know the "objective value" of anything in the world, a theory of inference based upon probability was required. On this impasse the Fisher-Jeffreys exchange ended.

Nothing was settled by the exchange. Neither scientist seems to have convinced the other of anything. Jeffreys later expressed some embarrassment over the whole affair referring to it as a "rather unfortunate discussion, which was full of misunderstandings on both sides" (Jeffreys, 1961, p. 393). As far as I know, Fisher never referred to the exchange again. Yet both men had been forced to examine some of their basic assumptions about probability and inference, and some weaknesses in their foundational views -- or at least their expository and polemical skills -- had been revealed. At any rate, some of Fisher's later writing on the foundations of inference was different in tone and theme; in some respects it appeared closer to Jeffreys' positions in the exchange than to Fisher's own.

4. Epilogue

By the 1950's, Fisher no longer regarded the Bayesian position as a going concern in the world of inference (see page 49 of [CP 242]). However, a new threat to the logic of induction had arisen: the extreme frequentism of the Neyman-Wald school, which Fisher criticized in [CP 261, 272, 273, 282] and [SMSI]. In response to this new threat, Fisher reformulated his definition of probability, and the new definition incorporated elements which, in the exchange with Jeffreys, he had condemned.

In particular, Fisher's new probability did seem to describe states of knowledge, and the central idea of the theory -- the principle of nonrecognition of relevant subsets -- bore a close resemblance to the Principle of Nonsufficient Reason; see pages 263-65 of [CP 272]. In fact, Fisher characterized the condition that no relevant subset be recognizable as "a postulate of ignorance" and -- as did the Principle of Nonsufficient Reason for Jeffreys -- this postulate provided the basis for an assignment of equiprobability to the members of a "set, perhaps of propositions, perhaps of events".

Not surprisingly, the term "subjective" disappeared from Fisher's pejorative vocabulary. Instead, Fisher came to regard the "objective" existence of probability values as a somewhat rare event, and instead he played up the role of imagination in assigning probabilities, as in this criticism of the Neymanian view:

> The root of the difficulty of carrying over the idea [of repeated sampling] from the field of acceptance procedures to that of tests of significance is that, where acceptance procedures are appropriate, the source of supply has an objective reality, and the population of lots, of one or more, which would be successively chosen for examination is uniquely defined; whereas if we possess a unique sample on which significance tests are to be performed, there is always ... a multiplicity of populations to each of which we can legitimately regard our sample as belonging; so that the phrases "repeated sampling" from the same population does not enable us to determine which population is to be used to define the probability level, for no one of them has objective reality, all being products of the statistician's imagination.

This does not sound like the same scientist who decried Jeffreys' notion of probability as "subjective and psychological".

Certainly Fisher never subscribed to the "inverse probability" he condemned in the 1930s. Yet in some of his last papers, for example [CP 267, 272], he explored more ways of using Bayes Theorem with priors derived from a fiducial argument. So far as I know, these ideas have not been developed by anyone since Fisher's death. But they provide an interesting final note to Fisher's lifelong fascination with Bayesian inference.

References

Jaynes, E.T. (1974). "Probability Theory with Applications in Science and Engineering. A Series of Lectures," unpublished manuscript.

Jeffreys, H. (1931). Scientific Inference. Cambridge: Cambridge University Press.

Jeffreys, H. (1932). "On the Theory of Errors and Least Squares," Proceedings of the Royal Society of London, Series A, 138, 48-55.

Jeffreys, H. (1933). "Probability, Statistics, and the Theory of Errors," Proceedings of the Royal Society of London, Series A, 140, 523-535.

Jeffreys, H. (1934). "Probability and Scientific Method," Proceedings of the Royal Society of London, Series A, 146, 9-15.

Jeffreys, H. (1961). Theory of Probability, 3rd edition. Oxford: Claredon Press.

Ramsey, F.P. (1961). "Truth and Probability". Reprinted in Studies in Subjective Probability (H. Kyburg and H. Smokler, eds.), 61-92.

DISCRIMINANT ANALYSIS

Somesh Das Gupta

1. Introduction

Between 1936 and 1940 Fisher published four articles on statistical discriminant analysis, in the first of which [CP 138] he described and applied the linear discriminant function. Prior to Fisher the main emphasis of research in this area was on measures of difference between populations based on multiple measurements. In this presentation, I shall briefly summarize the early work from Karl Pearson up to Mahalanobis, and then describe the major contributions associated with Fisher. These include the derivation of the linear discriminant function and its relationship to regression in the two-sample case, extensions to s populations for $s > 2$, and a test of significance associated with the linear discriminant function.

An interesting feature of Fisher's work in this area was his use of analogy with regression methods in the development of significance tests, often giving correct analyses through incorrect reasoning.

The presentation ends with a bibliography of articles on classification and discrimination, ordered by date, from 1916 to 1954, together with three recent general references on these topics.

2. Early Developments

The early development of discriminant analysis before Fisher dealt primarily with measures of difference between populations based on sample moments or frequency tables, and ignored the correlation among different variates. In 1916, Karl Pearson used the between-population distance measure

$$\chi^2 = \sum_{i=1}^{p} nn' \frac{(f_i/n_i - f_i'/n_i')^2}{f_i + f_i'}$$

based on the frequencies f_i and f_i' in the ith cell corresponding to two samples of sizes n and n'. Tildesley (1921) discussed a version of Pearson's moment distance measure, called the "Coefficient of Racial Likeness" (CRL),

$$\text{CRL} = \frac{1}{p}\sum_{i=1}^{p} \frac{(M_i - M_i')^2}{\sigma_i^2/n_i + \sigma_i'^2/n_i'} - 1,$$

where M_i, σ_i^2 and n_i denote respectively the mean, variance and the sample size of the first population, and M_i', $\sigma_i'^2$, n_i' similarly correspond to the second population. When $\sigma_i = \sigma_i'$ the CRL simplifies to

$$\frac{1}{p}\sum_{i=1}^{p} \frac{n_i n_i'}{n_i + n_i'} \left\{\frac{M_i - M_i'}{\sigma_i}\right\}^2 - 1.$$

The probable error of the CRL is $0.6749/\sqrt{2/p}$. In 1926, Pearson proposed as an alternative to the CRL, the measure

$$\Sigma^2 = \frac{1}{p}\sum_{i=1}^{p} \frac{(M_i - M_i')^2}{\sigma_i^2/n_i + \sigma_i'^2/n_i'},$$

and computed the standard error of Σ^2 to be $\sqrt{2/p}\,\{1 - 1/(2p)\}$ under the hypothesis of no population difference. His comment on the use of CRL was: "Not of how far the two races or tribes are alike or divergent, but how far on the given data we can assert significant resemblance or divergence...".

Mahalanobis, in his work between 1927 and 1930, suggested the distance measure (divergence in means)

$$D^2 = \frac{1}{p}\sum_{i=1}^{p} \frac{(M_i - M_i')^2}{\sigma_i^2} - \frac{1}{p}(1/n_i + 1/n_i')$$

as a criterion for discrimination in the equal variance case. He also discussed the problem of divergence in variability, skewness and kurtosis. On the basis of the moments and large sample approximations for D^2, Mahalanobis claimed large-sample superiority of D^2 over Pearson's CRL. A second measure proposed by Mahalanobis was

$$D_2^2 = \frac{1}{p}\sum_{i=1}^{p} \frac{(M_i - M_i')^2}{s_i^2} - \frac{1}{p}\sum_{i=1}^{p} \frac{\sigma_i^2}{s_i^2}(1/n_i + 1/n_i').$$

Pearson did not approve of Mahalanobis's work.

In 1931, Hotelling originated the now-famous T^2 statistic for testing a hypothetical value of the mean μ of a multivariate normal distribution. Suppose that $X_1,\ldots,X_N$ are independent p-variate random vectors following the $N_p(\mu,\Sigma)$

distribution, and define the sample statistics

$$\bar{x} = \sum_{\alpha=1}^{N} x_\alpha/N, \quad A = [a_{ij}] = \sum_{\alpha=1}^{N} (x_\alpha-\bar{x})(x_\alpha-\bar{x})'.$$

Then Hotelling's statistic for comparing $\bar{x}$ and μ is

$$T^2 = (N-1)N(\bar{x}-\mu)'A^{-1}(\bar{x}-\mu),$$

which he proved is distributed as $(N-1)p/(N-p)$ times an $F_{p,N-p}$ variate. He also noted that $N(\bar{x}-\mu)'\Sigma^{-1}(\bar{x}-\mu)$ is distributed as a χ_p^2 variate. E.S. Pearson and Wilks (1933) subsequently developed a test of the equality of mean vectors of k bivariate normal distributions with the same dispersion matrix.

In his work on the generalized distance between distributions, Mahalanobis (1936) introduced the between-population measure

$$\Delta^2 = \frac{1}{p} (\mu_1-\mu_2)'\Sigma^{-1}(\mu_1-\mu_2). \tag{1}$$

Note the link to his statistic, D^2, which can be rewritten as:

$$D^2 = \frac{1}{p} (\bar{x}_1-\bar{x}_2)'\Sigma^{-1}(\bar{x}_1-\bar{x}_2) - (1/n_1 + 1/n_2),$$

with the studentized form

$$D^2 = \frac{1}{p} (\bar{x}_1-\bar{x}_2)'C^{-1}(\bar{x}_1-\bar{x}_2), \tag{2}$$

in which C is the standard pooled estimate of Σ.

One of Fisher's first contacts with discriminant problems was in connection with M. Barnard's (1935) work on the secular variation of Egyptian skull characteristics. The theoretical content of Barnard's paper was rather weak, but the paper presents an interesting method for selecting discriminatory characteristics and forming a "weighted compound of the characters which will show the maximum secular change". For skulls from four different time periods, with seven characters measured on each skull, Barnard first selected four characters by a step-wise t-statistic comparison of the first and last groups. The four groups were then reduced by regression of characteristic means on the time variable, so that each selected characteristic was replaced by a regression coefficient and its associated standard error. The discrimination problem was then analyzed in terms of these regression coefficients. For a recent discussion of some aspects of this application see Rao (1973, Section 8d.5).

3. Fisher's Linear Discriminant Function [CP 138]

Fisher's 1936 paper dealt mostly with results needed for the comparison of two species, illustrated on the now-famous iris data. The sample size for each species was 50 and 4 characters were measured on each specimen. The question posed by Fisher was "What linear function of the four characters will maximize the ratio of the difference between the specific means to the standard deviation within species?"

If we let S denote the pooled "within sample" sum of squares and products matrix, and $d = \bar{x}_1 - \bar{x}_2$, then the goal is to find that 4×1 vector λ such that the non-centrality $(\lambda'd)^2/\lambda'S\lambda$ of the compound $\lambda'x$ is maximized, which turns out to be proportional to $\lambda = S^{-1}d$. Therefore, the desired linear function is taken as $d'S^{-1}x$. From this point on, Fisher treats λ as a fixed, known vector, and goes on to say that "The average variance of the two species in respect of the compound measurements may be estimated by dividing $[D = d'S^{-1}d]$ by 95;[1] the variance of the difference between two means of 50 plants each, by dividing again by 25". In a rather loose explanation for the divisor 95 in estimating variance, Fisher says "Since, in addition to the specific means, we have used three adjustable ratios, the variation [of the compound $\lambda'x$] within species must contain only 95 degrees of freedom". There follows an analysis of variance table, reproduced here as Table 1, summarizing the breakdown of variation in terms of $D = d'S^{-1}d$.

Table 1. Analysis of Variance of Linear Discriminant

Source	df	SS	
Between species	4	27.74160	$25D^2$
Within species	95	1.05361	D
Total	99	28.79501	$D(1 + 25D)$

[1] Fisher used D here despite the different usage of D in equation (2).

Fisher also considers the problem of classification in this paper, noting that: "The ratio of the difference between the means of the chosen compound measurements to its standard error in individual plants is of interest also in relation to the probability of misclassification". A plant from species 2 is misclassified as species 1 if

$$\lambda'x - \lambda'\bar{x}_2 > \tfrac{1}{2}(\lambda'\bar{x}_1 - \lambda'\bar{x}_2),$$

that is if x is closed to species 2 in the Mahalanobis sense

$$(x-\bar{x}_2)'S^{-1}(x-\bar{x}_2) > (x-\bar{x}_1)'S^{-1}(x-\bar{x}_1).$$

The probability of misclassification is then computed by treating $(\lambda'x - \lambda'\bar{x}_2)/\sqrt{\lambda'S\lambda/95}$ as a standard normal variate with $\lambda = S^{-1}d$, the result being a probability of "less than three per million".

The analysis of variance in Table 1 suggests to Fisher the analogy with partial regression, where in general $(x_1,\ldots,x_p)$ explain variation in the dummy variable y defined to equal $n_2/(n_1+n_2)$ and $-n_1/(n_1+n_2)$ in samples 1 and 2 respectively. If we write $k = n_1n_2/(n_1+n_2)$, then the least squares normal equation for regression coefficient b in the model y at b'x is

$$(S + kdd')b = kd,$$

from which b is found to be proportional to $k\lambda/(1+kD)$, with $\lambda = S^{-1}d$. The standard analysis of variance table, Table 2, is then obtained, and Fisher asserts "In this method of presentation the appropriate allocation of the degrees of freedom is evident". There follows a formal significance test of the variance ratio, essentially an F-test, with 4 and 95 degrees of freedom. Thus, apparently through incorrect reasoning, Fisher arrives at the correct significance test. It may be worth noting that a very similar line of reasoning was used in Example 46.2 of [SMRW], but there led to an incorrect significance test.

Table 2. Regression Analysis of Variance for Discriminant Analysis of Iris Data

Source	df	SS	
Regression	4	24.0854	$k^2D/(1 + kD)$
Remainder	95	0.9146	$k/(1 + kD)$
Total	99	25.	k

The closing section of the paper deals with aspects of linear discrimination among three species of iris, including the use of histograms of discriminant scores, $\lambda'x$.

4. Fisher's Second and Third Papers on Discrimination [CP 155 and CP 163]

In his 1938 paper [CP 155] Fisher reviewed the 1936 work and related it to the contributions by Hotelling (1931) and Mahalanobis (1936 and earlier), noting also that "In a very brilliant research R.C. Bose and S.N. Roy [1938] have demonstrated the distribution of D^2 [meaning $p^{-1}d'S^{-1}d$ as in equation (2)] is...". Fisher notes the correctness of his earlier significance test from Table 2, pointing out the difference with the usual normal-theory regression, and citing Hotelling's work.

The second half of the paper contains an attempt to extend the earlier work to the case of $s > 2$ populations.

> Measuring distance, we naturally will ask whether one observed distance significantly exceeds another. Measuring direction, we shall likewise be led to test whether three or more populations are collinear, or coplanar...

The theoretical analysis focusses on orthogonal linear contrasts among the s population means, but the treatment is rather incomplete (and was revised by Fisher in the Collected Papers).

The 1939 paper [CP 163] contains two sections devoted to the null distribution of the latent roots θ of the "between samples" dispersion matrix relative to the "within samples" dispersion matrix, assuming both degrees of freedom exceed p-1. The detailed calculations are carried out for $p = 2$, while "for p variates, the general distribution of the p roots is, as might ... at this stage be expected...".

It is worth mentioning here that, independently, Hsu (1938) and Girschick (1939) also derived the simultaneous distribution of the p roots.

5. Fisher's Test of a Proposed Linear Discriminant [CP 175]

In his 1940 paper Fisher dealt with the test of adequacy of a proposed linear discriminant function, for the two-population problem. The development is carried out largely using the dummy-variable regression technique of the 1936 paper.

The general version of the ANOVA table (cf. Table 2) is

Table 3. Regression ANOVA for Linear Discriminant Analysis

Source	df	SS
Regression	p	kR^2
Remainder	n_1+n_2-p-1	$k(1-R^2)$
Total	n_1+n_2-1	k

where as before $k = n_1n_2/(n_1+n_2)$ and R^2, the squared multiple correlation, is $D/(1+kD)$ with $D = d'S^{-1}d$. For a proposed linear discriminant $\xi = \beta'x$, we wish to test whether or not the distance between the two populations based on ξ is the same as the distance based on x. Let r denote the "within samples" correlation between ξ and the optimum linear discriminant $d'S^{-1}x$. Then the regression sum of squares in Table 3 is broken down, by the usual partial regression or covariance technique, into

sum of squares due to [regression on ξ] prediction $= kR^2r^2/\{1-R^2(1-r^2)\}$

sum of squares due to additional [regression] information

$$= \frac{kR^2(1-R^2)(1-r^2)}{1-R^2(1-r^2)},$$

with df = 1 and p-1 respectively. The formal test for the adequacy of ξ is then a test of the "additional information" sum of squares, or, as Fisher says "... modification of Hotelling's $[T^2]$ test ... (i) reducing the number of degrees of freedom by unity, and (ii) substituting ... $R^2(1-r^2)$ for R^2 ...". The F statistic is thus

$$\frac{n_1+n_2-p-1}{p-1} \; \frac{R^2(1-r^2)}{1-R^2(1-r^2)} \tag{3}$$

with p-1 and n_1+n_2-p-1 degrees of freedom under the null hypothesis, providing a valid test by analogy with Hotelling's T^2 result.

A more rigorous treatment of the problem would be to transform the multivariate normal variable X into Y = BX, with first component $\xi = \beta'X$ and the remaining components Z uncorrelated with ξ. Let $\Delta^2(X)$ and $D^2(X)$ denote the Δ^2 and D^2 measures of equations (1) and (2) based on variate X. Then, since Mahalanobis's Δ^2 measure in equation (1) is additive over independent variables, the desired test is equivalent to the test of the hypothesis that $\Delta^2(X) - \Delta^2(\xi) = \Delta^2(Z) = 0$. This leads to the Hotelling T^2 test based on Z, reducing to Fisher's result on re-expression in terms of X. Specifically, the T^2 statistic is

$$\frac{n_1+n_2-p-1}{p-1} \; \frac{kD^2(X)-kD^2(\xi)}{1 + kD^2(\xi)} ,$$

identical to equation (3).

The final section of the paper examines the application of discriminant techniques to the analysis of two-dimensional contingency tables. Optimal scores for row and column categories are derived, and a proposed system of scores is tested. Fisher refers here to Hotelling's related work on canonical variates.

References and Bibliography

The following list of articles, ordered by date of publication, includes those referred to in this lecture and others of historical importance.

Early Period

Pearson, K. (1916). "On the Probability that Two Independent Distributions of Frequency Are Really from the Same Population," Biometrika, 8, 250-254.

Tildesley, M.L. (1921). "A First Study of the Burmese Skull," Biometrika, 13, 247-251.

Pearson, K. (1926). "On the Coefficient of Racial Likeness," Biometrika, 18, 105-117.

Morant, G.M. (1926). "A First Study of Craniology of England and Scotland from Neolithic to Early Historic Times, with Special Reference to Anglo-Saxon Skulls in London Museums," Biometrika, 18, 56-98.

Mahalanobis, P.C. (1927). "Analysis of Race Mixture in Bengal," Journal and Proceedings of the Asiatic Society of Bengal, 23, 301-333.

Mahalanobis, P.C. (1928). "A Statistical Study of the Chinese Head," Man in India, 8, 107-122.

Morant, G.M. (1928). "A Preliminary Classification of European Races Based on Cranial Measurements," Biometrika, 20, 301-375.

Pearson, K. (1928). "The Application of the Coefficient of Racial Likeness to Test the Character of Samples," Biometrika, 20, 294-300.

Romanovsky, V. (1928). "On the Criteria that Two Given Samples Belong to the Same Normal Population," Metron, 7, 3-46.

Mahalanobis, P.C. (1930). "On Tests and Measures of Group Divergence," Journal and Proceedings of the Asiatic Society of Bengal, 26, 541-588.

Hotelling, H. (1931). "The Generalization of Student's Ratio," The Annals of Mathematical Statistics, 2, 360-378.

Woo, T.L. and Morant, G.M. (1932). "A Preliminary Classification of Asiatic Races Based on Cranial Measurements," Biometrika, 24, 108-134.

Pearson, E.S. and Wilks, S.S. (1933). "Methods of Statistical Analysis Appropriate for Samples of Two Variables," Biometrika, 25, 353-378.

Meyer, H.A. and Deming, W.E. (1935). "On the Influence of Classification on the Determination of a Measurable Characteristic," Journal of the American Statistical Association, 30, 671-677.

Mahalanobis, P.C. (1936). "On the Generalized Distance in Statistics," Proceedings of the National Institute of Sciences of India, 2, 49-55.

Contemporaries of Fisher

Barnard, M.M. (1935). "The Secular Variations of Skull Characters in Four Series of Egyptian Skulls," Annals of Eugenics, 6, 352-371.

Bose, S.N. (1936). "On the Complete Moment Coefficients of the D^2 Statistic," Sankhyā, 2, 385-396.

Bose, R.C. (1936). "On the Exact Distribution and Moment Coefficients of the D^2 Statistic," Sankhyā, 2, 143-154. Corrected and modified paper: Sankhyā, 2, 379-384.

Martin, E.S. (1936). "A Study of the Egyptian Series of Mandibles with Special Reference to Mathematical Methods of Sexing," Biometrika, 28, 149-178.

Smith, H.F. (1936). "A Discriminant Function for Plant Selection," Annals of Eugenics, 7, 240-250.

Bose, S.N. (1937). "On the Moment-Coefficients of the D^2-Statistic, and Certain Integral and Differential Equations Connected with the Multivariate Normal Populations," Sankhyā, 3, 105-124.

Bose, R.C. and Roy, S.N. (1937). "On the Distribution of Fisher's Taxonomic Coefficient and Studentised D^2-Statistic," Science and Culture, 3, 335.

Bose, R.C. and Roy, S.N. (1938). "The Distribution of the Studentised D^2-Statistic," Sankhyā, 4, 19-38.

Hsu, P.L. (1938). "Notes on Hotellings Generalized T^2," The Annals of Mathematical Statistics, 9, 231-243.

Bartlett, M.S. (1939). "The Standard Errors of Discriminant Function Coefficients," Supplement to the Journal of the Royal Statistical Society, 6, 169-173.

Girshick, M.A. (1939). "On the Sampling Theory of Roots of Determinantal Equations," The Annals of Mathematical Statistics, 10, 203-224.

Hsu, P.L. (1939). "On the Distribution of Roots of Certain Determinantal Equations," Annals of Eugenics, 9, 250-258.

Roy, S.N. (1939). "A Note on the Distribution of the Studentised D^2-Statistic," Sankhyā, 4, 373-380.

Roy, S.N. (1939). "P-Statistics and Some Generalizations in Analysis of Variance," Sankhyā, 4, 381-396.

Welch, B.L. (1939). "Note on Discriminant Functions," Biometrika, 31, 218-220.

Roy, S.N. and Bose, R.C. (1940). "The Use and Distribution of the Studentised D^2-Statistic when the Variances and Covariances are Based on Samples," Sankhyā, 4, 535-542.

Bhattacharya, D.P. and Narayan, R.D. (1941). "Moments of the D^2-Statistic for Populations with Unequal Dispersions," Sankhyā, 5, 401-412.

Goodwin, C.N. and Morant, G.M. (1941). "The Human Remains of Iron Age and Other Periods from Maiden Castle, Dorset," Biometrika, 31, 295-312.

Wald, A. (1944). "On a Statistical Problem Arising in the Classification of an Individual into One of Two Groups," The Annals of Mathematical Statistics, 15, 145-162.

Von Mises, R. (1954). "On the Classification of Observation Data into Distinct Groups," The Annals of Mathematical Statistics, 16, 68-73.

Recent References

Cacoullos, T. and Styan, G.P.A. (1973). "A Bibliography of Discriminant Analysis," in Discriminant Analysis and Applications edited by T. Cacoullos. New York: Academic Press.

Das Gupta, S. (1973). "Theories and Methods in Classification: A Review," in Discriminant Analysis and Applications edited by T. Cacoullos. New York: Academic Press.

Rao, C.R. (1973). Linear Statistical Inference and Its Applications. 2nd edition. New York: John Wiley and Sons. [Discussion of discriminant analysis and classification in Chapter 8].

This research was supported by U.S. Army Research Office Grant DAAG29-76-G-0038 to the University of Minnesota.

DISTRIBUTION ON THE SPHERE

Christopher Bingham

1. Introduction

The two primary goals of Fisher's 1953 paper "Dispersion on a Sphere" [CP 249] are: (i) to explore methodology for the analysis of more or less widely dispersed measurements of direction such as frequently arise in geology; and (ii) to "exhibit in a clear light" the "nature of inductive inferences", specifically to provide a non-standard example illustrating the correct application of fiducial inference.

A measurement of direction in either 2- or 3-dimensional space may clearly be identified with a point $\underset{\sim}{x}$ on the unit circle S_2 or unit sphere S_3, i.e., with a unit vector $\underset{\sim}{x}$. In two dimensions, the direction is naturally parameterized by the angle $\theta \in [0,2\pi)$ between the vector $\underset{\sim}{x}$ and an arbitrary direction. Thus $\underset{\sim}{x}' = (\cos\theta, \sin\theta)$. In three dimensions, two angles are needed. Probably most convenient are the colatitude $\theta \in [0,\pi]$ of $\underset{\sim}{x}$ relative to a fixed direction or pole (usually the vertical), and the longitude $\varphi \in [0,2\pi)$ "east" of a fixed direction orthogonal to the pole. With these coordinates we can express a unit vector as $\underset{\sim}{x}' = (\sin\theta\cos\varphi, \sin\theta\sin\varphi, \cos\theta)$. These 2- and 3-dimensional coordinate systems are illustrated in Figures 1(a) and 1(b), respectively.

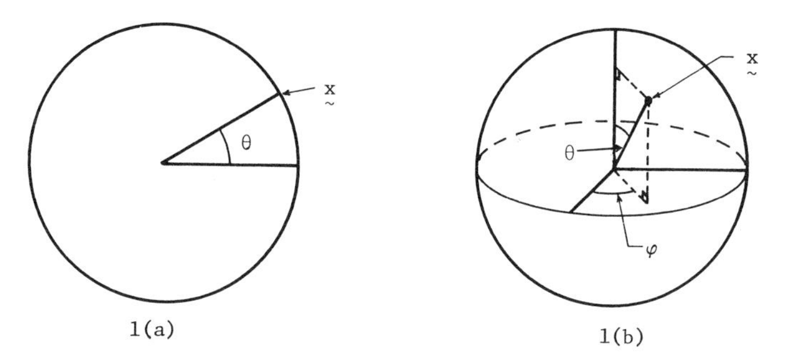

Figure 1: Polar Coordinates in Two and Three Dimensions

The only probability measure on the circle or sphere that is invariant under rotation is the uniform or isotropic distribution, namely $dS/2\pi = d\theta/2\pi$ for two dimensions and $dS/4\theta = \sin\theta\, d\theta\, d\varphi/4\pi$ for three dimensions, where dS represents the uniform Lebesgue measure. An alternative convenient expression for the isotropic distribution on the sphere is $dS/4\pi = (d\varphi/2\pi)(dc/2)$, where $c = \cos\theta = x_3$, which is a product of two independent uniform distributions on $[0,2\pi)$ and $[-1,+1]$, respectively.

2. The von Mises Distribution on the Circle

In the early part of this century, Richard von Mises (1918) considered the table of the atomic weights of elements, seven entries of which are as follows:

Element		Al	Sb	Ar	As	Ba	Be	Bi
Atomic Weight	W	26.98	121.76	39.93	74.91	137.36	9.01	209.00

He asked the question "Does a typical element in some sense have integer atomic weight?" A natural interpretation of the question is "Do the fractional parts of the weights cluster near 0 and 1?" The atomic weights W can be identified in a natural way with points on the unit circle, in such a way that equal fractional parts correspond to identical points. This can be done under the mapping

$$W \to \underset{\sim}{x} = \begin{bmatrix} \cos\theta_1 \\ \sin\theta_1 \end{bmatrix}, \quad \theta_1 = 2\pi(W-[W]),$$

where $[u]$ is the largest integer not greater than u. Von Mises' question can now be seen to be equivalent to asking "Do these points on the circle cluster near $\underset{\sim}{e}_1 = \begin{bmatrix} 1 \\ 0 \end{bmatrix}$?"

Incidentally, the mapping $W \to \underset{\sim}{x}$ can be made in another way:

$$W \to \underset{\sim}{x} = \begin{bmatrix} \cos\theta_2 \\ \sin\theta_2 \end{bmatrix}, \quad \theta_2 = 2\pi(W - [W+\tfrac{1}{2}]).$$

The two sets of angles for the two mappings are then as follows:

Element	Al	Sb	Ar	As	Ba	Be	Bi	Average
$\theta_1/2\pi$	0.98	0.76	0.94	0.91	0.36	0.01	0.00	$\bar{\theta}_1/2\pi = 0.566$
$\theta_2/2\pi$	-0.02	-0.24	-0.06	-0.09	0.36	0.01	0.00	$\bar{\theta}_2/2\pi = 0.006$

We note from the discrepancy between the averages in the final column that our usual ways of describing data, e.g., means and standard deviations, are likely to fail us when it comes to measurements of direction.

If the points do cluster near $\underset{\sim}{e}_1$ then the resultant vector $\sum_{j=1}^{N} \underset{\sim}{x}_j$ (here N = 7) should point in that direction, i.e., we should have approximately $\bar{\underset{\sim}{x}}/\|\bar{\underset{\sim}{x}}\| \doteq \underset{\sim}{e}_1$, where $\bar{\underset{\sim}{x}} = \Sigma\, \underset{\sim}{x}_j/N$ and $\|\underset{\sim}{x}\| = (\underset{\sim}{x}'\underset{\sim}{x})^{\frac{1}{2}}$ is the length of $\underset{\sim}{x}$. For the seven elements whose weights are considered here, we find

$$\bar{\underset{\sim}{x}}/\|\bar{\underset{\sim}{x}}\| = \begin{bmatrix} 0.9617 \\ -0.2741 \end{bmatrix} = \begin{bmatrix} \cos 344.09^\circ \\ \sin 344.09^\circ \end{bmatrix} = \begin{bmatrix} \cos -15.91^\circ \\ \sin -15.91^\circ \end{bmatrix} ,$$

a direction not far removed from $\underset{\sim}{e}_1$.

Von Mises then asked "For what distribution on the unit circle is the unit vector $\hat{\underset{\sim}{\mu}} = \begin{bmatrix} \cos \hat{\theta}_0 \\ \sin \hat{\theta}_0 \end{bmatrix} = \bar{\underset{\sim}{x}}/\|\bar{\underset{\sim}{x}}\|$ a maximum likelihood estimator (MLE) of a direction θ_0 of clustering or concentration?" The answer is the distribution now known as the von Mises or circular normal distribution. It has density, expressed in terms of random angle θ,

$$\frac{\exp\{k \cos (\theta-\theta_0)\}}{I_0(k)} \frac{d\theta}{2\pi},$$

where θ_0 is the direction of concentration and the normalizing constant $I_0(k)$ is a Bessel function. An alternative expression is

$$\frac{\exp(k\underset{\sim}{x}'\underset{\sim}{\mu})}{I_0(k)} \frac{dS}{2\pi}, \quad \underset{\sim}{\mu} = \begin{bmatrix} \cos \theta_0 \\ \sin \theta_0 \end{bmatrix}, \quad \|\underset{\sim}{x}\| = 1.$$

Von Mises' question clearly has to do with the truth of the hypothesis

$$H_0 : \theta_0 = 0 \quad \text{or} \quad \underset{\sim}{\mu} = \underset{\sim}{e}_1 = \begin{bmatrix} 1 \\ 0 \end{bmatrix}.$$

It is worth mentioning that Fisher found the same distribution in another context [SMSI, pp. 133-138] as the conditional distribution of $\underset{\sim}{x}$, given $\|\underset{\sim}{x}\| = 1$, when $\underset{\sim}{x}$ is $N_2(\underset{\sim}{\mu}, k^{-1}I_2)$.

3. The Fisher-von Mises Distribution

Fisher's 1953 paper [CP 249] is about the analogue of the von Mises distribution on the sphere. If $\underset{\sim}{x}$ is tri-variate normal $N_3(\underset{\sim}{\mu}, k^{-1}I_3)$, $\|\underset{\sim}{\mu}\| = 1$, then its conditional distribution given $\|\underset{\sim}{x}\| = 1$, i.e., given that $\underset{\sim}{x}$ lies on the unit sphere, is

$$\frac{\exp(k\underset{\sim}{x}'\underset{\sim}{\mu})}{F(k)} \frac{dS}{4\pi}, \quad \underset{\sim}{x} \in S_3, \; \underset{\sim}{\mu} \in S_3, \; k \geq 0, \tag{1}$$

with $F(k) = \frac{1}{2} \int_0^\pi \exp(k \cos \theta) \sin \theta \, d\theta = k^{-1} \sinh k$, or, in terms of the angular distance θ between $\underset{\sim}{x}$ and $\underset{\sim}{\mu}$, by

$$\frac{\exp(k \cos \theta)}{F(k)} \frac{dS}{4\pi}, \quad \theta \in [0,\pi].$$

Yet another way of expressing this distribution is as follows: Let $\delta = 2 \sin (\theta/2)$ be the straight line distance between the points $\underset{\sim}{x}$ and $\underset{\sim}{\mu}$, and let φ represent "longitude" with $\underset{\sim}{\mu}$ as "north" pole. Then $\cos \theta = 1 - \delta^2/2$ and $dS = \delta \, d\delta \, d\varphi$, and the distribution (1) can be expressed as

$$\frac{\exp(k)}{2F(k)} \delta \exp(-k\delta^2/2) d\delta \frac{d\varphi}{2\pi}, \quad \delta \in [0,2], \; \varphi \in [0,2\Pi)$$

$$= \{1 - \exp(-2k)\}^{-1} (k/2) \; \delta \exp(-k\delta^2/2) d\delta \frac{d\varphi}{2\pi}.$$

The distribution of $k\delta^2$ is that of $\chi^2_{(2)}$, truncated at 4k. Moreover, for large k, the projection of $\underset{\sim}{x}$ on a plane orthogonal to $\underset{\sim}{\mu}$ is approximately $N_2(0,k^{-1}I_2)$. This says that for large k we can approximate the Fisher-von Mises distribution by a bivariate normal distribution in the tangent plane at $\underset{\sim}{\mu}$, centered at $\underset{\sim}{\mu}$ with covariance matrix $k^{-1}I_2$. Thus, for highly concentrated samples, we can ignore the fact that we are dealing with data on the sphere and apply standard statistical methodology. This is the case, for example, if we are trying to combine several accurate measurements of the location of a star in the celestial sphere.

The log likelihood function based on a sample $\underset{\sim}{x}_1,\ldots,\underset{\sim}{x}_N$ is, from expression (1),

$$N\{-\log F(k) + k\bar{\underset{\sim}{x}}'\underset{\sim}{\mu}\}, \qquad \bar{\underset{\sim}{x}} = N^{-1}\sum_{j=1}^{N} \underset{\sim}{x}_j .$$

We can see immediately that $\bar{\underset{\sim}{x}}$ is sufficient for $\underset{\sim}{\mu}$ and k and that the MLE of $\underset{\sim}{\mu}$ is $\hat{\underset{\sim}{\mu}} = \bar{\underset{\sim}{x}}/||\underset{\sim}{x}||$. Moreover the MLE of k is the unique solution of

$$(d/dk)\ \log F(k) = \bar{\underset{\sim}{x}}'\hat{\underset{\sim}{\mu}} = ||\bar{\underset{\sim}{x}}|| = ||\Sigma\ \underset{\sim}{x}_j||/N = R/N,$$

where R is the length of the resultant vector $\Sigma\ \underset{\sim}{x}_j$. If $\underset{\sim}{\mu}$ is known, then $X \equiv \Sigma\ \underset{\sim}{x}_j'\underset{\sim}{\mu} = RC$, where $C = \bar{\underset{\sim}{x}}'\underset{\sim}{\mu}/||\bar{\underset{\sim}{x}}||$ is the cosine of the angle between $\bar{\underset{\sim}{x}}$ and $\underset{\sim}{\mu}$, is sufficient for k. The MLE of k in this case is the solution of (d/dk) log F(k) = X/N. Also note that $2(N-X) = \sum_{j=1}^{N} \delta_j^2$ is approximately distributed as $k^{-1}\chi^2_{2N}$ for large k.

4. Sampling Distributions

Let $g_N(\bar{\underset{\sim}{x}};k,\underset{\sim}{\mu})d\bar{\underset{\sim}{x}}$ represent the sampling distribution of $\bar{\underset{\sim}{x}}$, $N \geq 2$. That is

$$g_N(\bar{\underset{\sim}{x}};k,\underset{\sim}{\mu})d\bar{\underset{\sim}{x}} = \int_{\bar{x}\ \mathrm{const}}\cdots\int \frac{\exp(Nk\bar{\underset{\sim}{x}}'\underset{\sim}{\mu})}{F^N(k)}\frac{dS_1\ \cdots\ dS_N}{(4\pi)^N}$$

$$= \frac{\exp(Nk\bar{\underset{\sim}{x}}'\underset{\sim}{\mu})}{F^N(k)}\int_{\bar{x}\ \mathrm{const}}\cdots\int \frac{dS_1\ \cdots\ dS_N}{(4\pi)^N}$$

$$= \frac{\exp(Nk\bar{\underset{\sim}{x}}'\underset{\sim}{\mu})}{F^N(k)}\, g_N(\bar{\underset{\sim}{x}};0,\underset{\sim}{\mu})d\bar{\underset{\sim}{x}},$$

where $g_N(\bar{\underset{\sim}{x}};0,\underset{\sim}{\mu})$ is the density of $\bar{\underset{\sim}{x}}$ when k = 0, i.e., when the $\underset{\sim}{x}_i$'s are distributed uniformly on the sphere. Thus hard work needs to be done only for the uniform case. It is clear that in this case the direction $\hat{\underset{\sim}{\mu}} = \bar{\underset{\sim}{x}}/||\bar{\underset{\sim}{x}}||$ of $\bar{\underset{\sim}{x}}$ is itself uniform on the sphere and is independent of $||\bar{\underset{\sim}{x}}|| = R/N$. Thus, it follows that when k = 0, the distribution of $\bar{\underset{\sim}{x}}$ must be of the form

$$g_N(\bar{\underset{\sim}{x}};0;\underset{\sim}{\mu})d\bar{\underset{\sim}{x}} = f_R(R)dR\,\frac{dC}{2}\,\frac{d\varphi}{2\pi},\quad N \geq 2,$$

where f_R is the density of R and φ is "longitude" of $\hat{\underset{\sim}{\mu}}$ around $\underset{\sim}{\mu}$. Clearly φ is irrelevant to any inference about $\underset{\sim}{\mu}$ and k, and hence we need deal only with the joint density of R and C.

Fisher shows (see Lord (1954) for more general results) that

$$f_R(R) = -2R(d/dx)f_X(x)\Big|_{x=R},$$

where f_X is the density of $X = \sum_{j=1}^{N} \cos\theta_j$, $\cos\theta_j = \underset{\sim}{x}_j'\underset{\sim}{\mu}$. He also derives the known result that X, being the sum of M independent uniform variables on $[-1,+1]$, has density

$$f_X(x) = 2^{-N}\{(N-1)!\}^{-1} \sum_{2j\leq N-x} (-1)^j \binom{N}{j}(N-x-2j)^{N-1},\quad |x| \leq N.$$

Defining

$$\phi_N(x) = -2^N(d/dx)f_X(x) = \{(N-2)!\}^{-1} \sum_{2j\leq N-x} (-1)^j\binom{N}{j}(N-x-2j)^{N-2},\quad |x| \leq N,$$

we have

$$f_R(R) = R\phi_N(R)/2^{N-1},\qquad 0 \leq R \leq N,$$

and the joint distribution of R and C is, when $k = 0$,

$$2^{-N}\phi_N(R)\,RdRdC.$$

Finally, when $k > 0$, the joint distribution of R and C is, since $N\bar{\underset{\sim}{x}}'\underset{\sim}{\mu} = RC$,

$$h_N(R,C;k)dRdC = \frac{\exp(kRC)}{F^N(k)}\,\frac{\phi_N(R)}{2^N}\,RdRdC,\ C \in [-1,1],\ R \in [0,N].$$

The marginal distribution of R is obtained by integrating out C from $h_N(R,C,;k)$. This yields

$$f_R(R;k)dR = \frac{2F(kR)}{\{2F(k)\}^N}\,\phi_N(R)dR$$

$$\cong 2^{N-1}\{1 + (R^2 - N)k^2/6 + O(k^4)\}\phi_N(R)dR, \tag{2}$$

for small k. Finally, the conditional distribution of C given R is

$$f(C|R)dC = \frac{\exp(kRC)}{F(kR)} \frac{dC}{2} = \{1 - \exp(-2kR)\}^{-1} kR \exp\{kR(C-1)\} dC \tag{3}$$

which implies that conditionally given R, $\hat{\underset{\sim}{\mu}}$ has the Fisher-von Mises distribution centered at $\underset{\sim}{\mu}$ with concentration parameter kR.

Incidentally, Fisher seems to imply that the appropriate MLE for k should be based on the marginal distribution of R. Equation (2) shows that this has the appealing property that when R is very small, specifically $R < N^{\frac{1}{2}}$, the MLE of k is 0, whereas the MLE of k based on the joint distribution of R and $\hat{\underset{\sim}{\mu}}$ is always positive unless R = 0.

5. Fiducial Inference According to Fisher

In Fisher's opinion, inference about $\underset{\sim}{\mu}$ should be based on the conditional distribution (3) of $C = \hat{\underset{\sim}{\mu}}'\underset{\sim}{\mu}$ given R. If R were sufficient for k in the standard meaning of the term, then this conditional distribution would not involve k and we could get exact small sample inference. Fisher proceeds to eliminate the nuisance parameter k "by the method first proposed by 'Student'," namely, he obtains its fiducial distribution given R and then integrates k out from the conditional distribution of C given R. We can summarize these steps as follows:

1. Find the fiducial distribution of $\hat{\underset{\sim}{\mu}}$, or equivalently of C, conditional on R, for fixed k.
2. Find the fiducial distribution of k based on the marginal distribution of R.
3. Integrate out k from the result of step 1 using the result of step 2.

This procedure is similar to one used in getting the fiducial distribution of the mean μ in a normal population. There the analogous steps are:

1. Find the fiducial distribution of μ, conditional on s^2 for fixed σ^2; this is $N_1(\bar{x},\sigma^2/N)$.
2. Find the fiducial distribution of σ^2; this is $\sigma^2 \sim (N-1)s^2/\chi^2_{N-1}$.
3. Integrate out σ^2 from the result of step 1 using the result of step 2. This yields $\mu \sim \bar{x} + N^{-\frac{1}{2}} s t_{N-1}$, the usual rescaled "Student's" t distribution.

Now for large k, the distribution of R is approximately $k^{N-1}\exp\{k(N-R)\}$ $\phi_N(R)dR$. If $R > N-2$ (implying the MLE $\hat{k} > (N-1)/2$), then $\phi_N(R) = (N-R)^{N-2}/(N-2)!$, and thus the marginal distribution of R is, in this range,

$$\{(N-2)!\}^{-1}k^{N-1}(N-R)^{N-2}\ \exp(-k(N-R)dR,$$

except for an omitted term involving exp(-2k). Thus the fiducial distribution of k is

$$k \sim \chi^2_{2N-2}/2(N-R), \qquad R \geq N-2.$$

Finally, for large k, the distribution of C given R is, neglecting a term of magnitude exp(-kR), $kR\exp\{kR(C-1)\}dC$. Hence the fiducial distribution of C for $R \geq N-2$ is

$$(N-R)^{N-1}[\int_0^\infty \frac{k^{N-2}}{(N-2)!}\ \exp\{-k(N-R)\}\ kR\exp\{-kR(1-C)\}dk]dC$$

$$= (N-1)\ \frac{(N-R)^{N-1}}{(N-RC)^N}\ RdC = (N-1)\ \frac{(N-R)^{N-1}}{(N-X)^N}\ RdC.$$

Integrating this with respect to C from $-\infty$ to C_0 (not the appropriate -1 to C_0) we obtain the fiducial probability statement for $\underset{\sim}{\mu}$, when $\hat{\underset{\sim}{\mu}}$ and R have been observed, as

$$P = P(\hat{\underset{\sim}{\mu}}'\underset{\sim}{\mu} \leq C_0) = \{(N-R)/(N-RC_0)\}^{N-1}, \quad R \geq N-2. \tag{4}$$

According to Fisher, the general result (still assuming R is large enough so that exp(-2kR) can be ignored in the conditional distribution of C given R) is

$$P = \sum_{2j \leq N-R} (-1)^j \binom{N}{j} \frac{R(1-C)}{R(1-C)+2j}\ \{(N-R-2j)/(N-RC)\}^{N-1}. \tag{5}$$

A proof or disproof of expression (5) would be welcome, since Fisher never gave a proof nor have I seen one by anyone else.

6. The Direct Method According to Watson and Williams

Watson and Williams (1956) developed an alternative "direct" approach to inference on $\underset{\sim}{\mu}$ which Williams (1963) put in direct contrast with Fisher's. The direct method follows a path analogous to one derivation of "Student's" t as a distribution appropriate to inference on μ in the usual normal case. As pointed out earlier, the random variable $X = (\sum_{j=1}^{N} \underset{\sim}{x}_j)'\underset{\sim}{\mu}$ is sufficient for k, in the standard sense that the conditional distribution of the sample (and in particular of R) given X does not depend on k. For any hypothesized value of $\underset{\sim}{\mu}$, X is known, and Watson and Williams propose that inference as to the truth of this hypothesis should be based on this conditional distribution. Since it does not depend on k, we can take the simplest case when $k = 0$. But then, as given above, the joint distribution of R and C is $2^{-N}\phi_N(R)RdRdC$, and hence the joint distribution of R and $X = RC$ is $2^{-N}\phi_N(R)dRdX$, with $|X| \leq R \leq N$. Then the conditional distribution, given X, of $R = X/C$ is $\dfrac{\phi_N(R)}{\chi_N(|X|)}\,dR$, where

$$\chi_N(t) = \int_t^N \phi_N(R)\,dR = \{(N-1)!\}^{-1} \sum_{2j\leq N-t} (-1)^j \binom{N}{j}(N-t-2j)^{N-1}.$$

The P-value of a test of $\underset{\sim}{\mu}$ based on this conditional distribution is

$$P = \chi_N(R)/\chi_N(|X|) = \frac{\sum_{2j\leq N-R} (-1)^j \binom{N}{j}(N-R-2j)^{N-1}}{\sum_{2j\leq N-|X|} (-1)^j \binom{N}{j}(N-|X|-2j)^{N-1}}$$

Note that if $N-2 \leq |X| \leq R \leq N$, then $P = \{(N-R)/(N-|X|)\}^{N-1}$, which coincides with Fisher's result (4) for $R \geq N-2$. From its derivation it can be seen that the test is "exact", requiring no assumption of large k. On the other hand, it has an important flaw: If $\hat{\underset{\sim}{\mu}}$ is near $-\underset{\sim}{\mu}$, i.e., almost opposite to the direction supposed by the null hypothesis, then we may be unable to reject the null hypothesis even for large R and N since the test depends on $|X|$ rather than X.

In this direct method, inference on $\underset{\sim}{\mu}$ is reached via two steps:

1. Find the conditional distribution of $\hat{\underset{\sim}{\mu}}$ given the sufficient quantity X for a hypothesized value of $\underset{\sim}{\mu}$.
2. Compute the conditional probability of a value for $\hat{\underset{\sim}{\mu}}$ to be more deviant from the hypothesized $\underset{\sim}{\mu}$ than the observed $\hat{\underset{\sim}{\mu}}$. If this is small enough reject the null hypothesis.

The set of $\underset{\sim}{\mu}$'s not rejected at level α by this procedure is an "exact" 1-α confidence region for $\underset{\sim}{\mu}$. It will, however, always have antipodal symmetry, including directions near $-\hat{\underset{\sim}{\mu}}$, as well as those near $\hat{\underset{\sim}{\mu}}$. If one discards the half in the opposite hemisphere, the "exactness" is lost.

The analogous direct method leading to "Student's" t has the following steps:

1. Find the conditional distribution of $\hat{\mu} = \bar{x}$ given the sufficient quantity $S = \sum_{i=1}^{N} (x_j-\mu)^2$ for a hypothesized value of μ. This yields $\bar{x} - \mu \sim (S/N)^{\frac{1}{2}}(2v-1)$, where v has a beta distribution with both parameters equal to $\frac{1}{2}(N-1)$.
2. Compute the conditional probability of a value for $\bar{x}$ more deviant from the hypothesized μ than the observed $\bar{x}$. If this is small enough reject the null hypothesis.

Inverting this test yields the usual confidence interval for μ based on "Student's" t.

7. Later Discussion of Inference

On the whole, it appears that Fisher failed in the second of the goals stated for his 1953 paper [CP 249]. Although the Fisher-von Mises distribution has become a principal tool for analyzing spherical data (see Mardia, 1972, 1975; Watson, 1956; Irving, 1964, among many others), its role as a non-standard example of fiducial inference has not received comparable attention. Williams (1963) and Barnard (1963) in companion papers were the first explicitly to contrast the two approaches. Williams, seemingly in the tradition of Fisher, argues for the importance of sufficiency, emphasizing that the direct method is based on the sufficiency for k of the quantity $X = X(\underset{\sim}{\mu}) = R\hat{\underset{\sim}{\mu}}'\underset{\sim}{\mu}$, which, of course, differs for each hypothesized $\underset{\sim}{\mu}$. Barnard, in contrast, claims that the conventional definition of sufficiency used by Williams does not really reflect Fisher's conception. Quoting Barnard, "Fisher's definition of sufficiency was designed to embody a logical notion, that of providing the whole of the available relevant information for a given parameter" (emphasis Barnard's). It seems clear intuitively that when nothing is known about $\underset{\sim}{\mu}$, all the information concerning k available from the data is subsumed in R, the length of the resultant vector. Barnard makes precise this observation by defining the notion of "sufficient for θ in the absence of knowledge of γ," using aspects of group invariance. He does not, however, discuss what to me is the weakest point in Fisher's derivation, namely integrating k out of the conditional distribution of $\underset{\sim}{\mu}$ given R using the "fiducial" distribution of k. In a Bayesian approach to the problem, Mardia and El-Atoum (1976) suppose the improper prior $k^{-1}dkdS(\underset{\sim}{\mu})/4\pi$. For large R this yields Fisher's solution as the posterior distribution of $\underset{\sim}{\mu}$. A discussion of this problem that is closely related to Barnard's can be found in Fraser (1968, pp. 196-202 and 220-223) where the concept of "structural inference" is applied to inference concerning $\underset{\sim}{\mu}$.

References

Barnard, G.A. (1963). "Some Logical Aspects of the Fiducial Argument," Journal of the Royal Statistical Society, Series B, 25, 111-114.

Fraser, D.A.S. (1968). The Structure of Inference. New York: John Wiley and Sons.

Irving, E. (1964). Paleomagnetism, and Its Application to Geological and Geophysical Problems. New York: John Wiley and Sons.

Lord, R.D. (1954). "The Use of the Hankel Transform in Statistics, I. General Theory and Examples," Biometrika, 41, 44-55.

Mardia, K.V. (1972). Statistics of Directional Data. London and New York: Academic Press.

Mardia, K.V. (1975). "Statistics of Directional Data (with Discussion)," Journal of the Royal Statistical Society, Series B, 37, 349-393.

Mardia, K.V. and S.A.M. El-Atoum (1976). "Bayesian Inference for the von Mises-Fisher Distribution," Biometrika, 63, 203-206.

von Mises, R. (1918). "Uber die 'Ganzzahligheit' der Atomgewichte und Verwandte Fragen," Physikalische Zeitschrift, 7, 153-159.

Watson, G.S. (1956). "Analysis of Dispersion on a Sphere," Monthly Notices of the Royal Astronomical Society, Geophysical Supplement, 7, 153-159.

Watson, G.S. and E.J. Williams (1956). "On the Construction of Significance Tests on the Circle and the Sphere," Biometrika, 43, 344-352.

Williams, E.J. (1963). "A Comparison of the Direct and Fiducial Arguments in the Estimation of a Parameter," Journal of the Royal Statistical Society, Series B, 25, 95-99.

SMOKING AND LUNG CANCER

R. Dennis Cook

1. Introduction

During the years 1957 and 1958, Fisher became involved in a controversy over the issue of whether or not it had been established that smoking causes lung cancer. The seeds of the controversy were planted in a series of reports by Doll and Hill in the British Medical Journal, which showed a significant association between the incidence of smoking and the incidence of lung cancer in a retrospective study. There ensued, in 1957, much alarum in the eaitorial pages of the British Medical Journal, which Fisher felt to be unjustified, in the sense that hard scientific evidence of causation was yet to be found. In a series of letters, exchanges, and lectures, Fisher attempted to criticize the weakness of the evidence for causation, and to propose possible explanations for the observed effects which needed to be investigated. He also pointed to a seemingly absurd aspect of Doll and Hill's data. This lecture reviews Fisher's participation in the controversy.

2. The Doll and Hill Study

During 1948-49, Doll and Hill obtained data on 709 lung cancer patients[1] in 20 London hospitals. For comparative purposes a control group of 709 patients without lung cancer was obtained, having essentially the same distribution with regard to sex, age, hospital location and so forth. Persons were then classified as to their smoking habits, "smoker" being defined as one who smoked at least once a day for a year. Data obtained included answers to the following questions: Smoker or not? Starting and stopping ages? Maximum ever smoked? Pipes or cigarettes? Inhale smoke or not? Also obtained were age, sex, social class and place of residence. A second interview of 50 unselected control patients was

[1] These data form a subset of the whole study.

carried out as a check on possible imbalance between the cancer and control groups. Partial results are reproduced here in Tables 1 and 2.

Table 1: Incidence of Smoking in the Cancer and Control Groups, 1948-49 Study

	Men		Women	
	Smoker	Non-Smoker	Smoker	Non-Smoker
Cancer	647	2 (0.3%)	41	19 (31.7%)
Control	622	27 (4.2%)	28	32 (53.5%)
Exact P-value = 0.64×10^{-6}			Chi-square = 5.76, P-value = 0.025	

Table 2: Incidence Rates of Types and Methods of Smoking in the Cancer and Control Groups

	(a) Male Smokers		
	Pipe Only	Cigarettes Only	Total
Cancer	5.7%	94.3%	525
Control	9.7%	90.3%	507
Chi-square = 5.70			

	(b) All Smokers (Men and Women)		
	Inhaler	Non-Inhaler	Total
Cancer	61.6%	38.4%	688
Control	67.2%	32.8%	650
Chi-square = 4.58			

The data on inhaling, which might be used to suggest that inhalers are less prone to lung cancer than non-inhalers, played an important role in Fisher's discussions. In 1952, Doll and Hill reported additional data including the results in Table 3. Table 4 combines the results of both studies on the question of inhaling.

Table 3: Incidence Rates of Inhalers Among Smokers in Cancer and Control Groups, 1952 Data

	Inhaler	Non-Inhaler	Total
Cancer	67.5%	32.5%	727
Control	65.9%	34.1%	693
Chi-square = 0.41			

Table 4: Incidence Rates of Inhalers, 1948-49 and 1952 Data (Tables 2b and 3) Combined

	Inhaler	Non-Inhaler	Total
Cancer	64.6%	35.4%	1415
Control	66.6%	33.4%	1343

Chi-square = 1.18

3. Correspondence in the British Medical Journal

The last of three reports by Doll and Hill appeared in 1956. On June 29, 1957, an editorial in the British Medical Journal contained a strong statement on the cause and effect interpretation of smoking and lung cancer. Three quotes from the editorial page are of particular interest:

> ... Sir Alfred Egerton, F.R.S. and Sir Ronald Fisher, F.R.S., have agreed to become [the Imperial Tobacco Company's] scientific consultants...

> ... In the presence of the painstaking investigations of statisticians that seem to have closed every loophole of escape for tobacco as the villain in the piece, it is well to remember...

> ... But the hazards to health and life are undeniable, hazards which must be brought home to the public by all the modern devices of publicity...

Fisher did not like this, especially the last quote. He felt that the case must be analyzed in a scientific forum rather than in the public press, and more research must be done before a sound conclusion could be drawn. He expressed his objections in a letter to the British Medical Journal of July 6, 1957 [CP 269]:

> ... In recent wars, for example, we have seen how unscrupulously the "modern devices of publicity" are liable to be used under the impulsion of fear; and surely the "yellow peril" of modern times is not the mild and soothing weed but the organized creation of states of frantic alarm...

> ... Many would still feel, as I did about five years ago, that a good prima facie case had been made for further investigation. None think that the matter is already settled. The further investigation seems, however, to have degenerated into the making of more confident exclamation, with the studied avoidance of the discussion of those alternative explanations of the facts which still await exclusion...

At the end of the letter, Fisher asked: "Is not the matter serious enough to require serious treatment?"

Two letters to the Editor of the British Medical Journal on July 20, 1957 responded to Fisher's objections:

(1) Robert McCurdy's Letter: McCurdy began his letter by saying: "Sir Ronald Fisher's criticism of your leading article on this subject would not be so unfair if he had specified what alternative explanations of the facts still await exclusion". As for the two alternative explanations suggested by the Tobacco Manufacturer's Annual Report, namely, (i) genetic predisposition (GP) to lung cancer is associated with a genetic predisposition to smoke, and (ii) smoking acts in association with some other factor such as atmospheric pollution, McCurdy's responses can be summarized as: (a) GP is proportional to the amount smoked, (b) GP (cigarette) > GP (pipes), (c) GP has increased over the past 30 years, more rapidly in men than women, (d) GP (give up) < GP (continue to smoke), and (e) data do not support the second explanation above. McCurdy ended his letter by saying:

> ... The Medical Research Councils have said that the most reasonable interpretation of the evidence is that the relationship between tobacco-smoking, particularly in the form of cigarettes, and lung cancer is one of direct cause and effect. I suggest, Sir, that this is one of those masterly understatements for which we as a nation are famous...

(2) John Burton's Letter: Burton objected to Fisher's calling the use of publicity "propaganda". Being the Medical Director for the Central Council for Health Education, he felt obliged to inform the public of those "facts" unemotionally and as thoroughly as possible, i.e., those who smoke run a considerably greater risk of cancer of the lung than those who do not, cigarette-smoking involves a greater risk than smoking pipes or cigars, the risk increases with the number of cigarettes smoked, and that giving up smoking is effective in decreasing the risk.

Fisher's response to these two letters appeared in a letter to the Editor of the British Medical Journal on August 3, 1957 [CP 270] in which he proposed two alternative theories: (i) cancer causes cigarette smoking, and (ii) individual genotype causes both. His explanation was that logical procedures always allow three explanations: (i) A is the cause of B, (ii) B is the cause of A, and (iii) both A and B are caused by a third factor. Therefore, he proposed the obvious two alternative theories to "smoking causes cancer". The arguments in the letter contain the following points: (a) genetic theory cannot explain secular changes, but neither can any other theory, (b) genetic theory can explain the difference between pipe and cigarette smokers, (c) inbred strains of mice show differences in the frequency, age-incidence and type of cancer, and (d) in man, stomach cancer is associated with the gene for blood group A. Fisher also pointed out in the letter that excessive confidence in having found the solution is the main obstacle in the way of more penetrating research. As he said:

> ... We should not be content to be "not so unfair", for without fairness the statistician is in danger of scientific errors through his moral fault.... For it will be as clear in retrospect, as it is now in logic, that the data so far do not warrant the conclusions based on them....

4. Fisher's 1958 Lecture

In a lecture given at Michigan State University [CP 274], Fisher gave an expanded version of his earlier views on the association between smoking and lung cancer. He reiterated his objection to vehement publicity on the basis of *prima facie* evidence, and demanded that moral distinction and rational attitudes were necessary. There is a long discourse on the many possibilities that he believed to require investigation before the causes of lung cancer could be established. Fisher notes the impossibility of experimentation, and comments on the ethical questions surrounding randomized trials. He also admits that medical science has indeed made progress on the basis of purely observational evidence. Nevertheless, he is firm on the point that much more needs to be and can be done in the case of lung cancer. The lecture ends "And so I should like to see those two things done, one immediately and quickly: an inquiry into the effects on inhaling, and secondly, a more difficult but certainly a possible task of seeing to what extent different smoking classes were genotypically conditioned. And I believe that only overconfidence, if it is allowed to have its way, could prevent those further inquiries from being made".

5. Fisher's Numerical Evidence for the Genetic Theory

In 1958, Fisher obtained data on the association between genotype and smoking habit for monozygotic and dizygotic twins. The results are given here in Table 5 and regrouped in Table 6.

Table 5: Smoking Habits of Identical and Non-Identical Twins

	Non-smoker	Cigarette	Cigar	Somewhat Alike	Different	Total
Monozygotic	9	22	2	6	12	51
Dizygotic	0	11	0	4	16	31

Table 6: Brief Form of Table 5 Data

	Alike	Somewhat Alike	Different
Monozygotic	33 (65%)	6 (12%)	12 (23%)
Dizygotic	11 (35%)	4 (13%)	16 (52%)

In a letter to the Editor of Nature in 1958 [CP 275], Fisher commented that McCurdy had rejected the genetic theory with some contempt. As for the genetic theory, he points out that only when different smoking classes are genotypically homogeneous can causation be confidently inferred from the association observed, i.e., we must rule out, on the basis of data, other possibilities. In view of the results tabulated above, he said in his letter:

> ... There can therefore be little doubt that the genotype exercises a considerable influence on smoking, and on the particular habit of smoking adopted, and that a study of twins on a comparatively small scale is competent to demonstrate the rather considerable differences which must exist between the different groups who classify themselves as nonsmokers, or the different classes of smokers. Such genotypically different groups would be expected to differ in cancer incidence; and their existence helps to explain such oddities as that pipe and cigar smokers should show much less lung cancer than cigarette smokers, while among the latter, the practice of inhaling is associated with less rather than with more cancer of the lung...

However, Fisher's argument has the flaw that monozygotic twins can not be treated as individuals who behave independently of each other. This criticism was pursued by an anonymous writer "Geminus" in the New Scientist (1958). "Geminus" suggested that the correlation of psychological attitudes of identical twins "might easily account for Sir Ronald Fisher's results", and dismissed Fisher's work as a smoke-screen between heavy smokers "and what is rapidly becoming an accepted truth -- that smoking can cause lung cancer".

Fisher responded to "Geminus" in a second letter to Nature [CP 276], which refers to the continuing use of "modern publicity" and the use of eloquence in place of evidence. He restates the effect of inhaling and the value of the genetic theory; previous data providing positive prima facie evidence that the genotype is not unimportant. Further data had been obtained in London on female identical twins, some raised separately and the others raised together (see Tables 7 and 8).

Table 7: Smoking Habits of Female Twins

	Alike	Not Alike	Total
Monozygotic	44 (83%)	9 (17%)	53
Dizygotic	9 (50%)	9 (50%)	18

Table 8: Smoking Habits of Female Identical Twins

	Alike	Not Alike	Total
Raised Separately	23	4	27
Raised Together	21	5	26

In conclusion Fisher said:

> There is nothing to stop those who greatly desire it from believing that lung cancer is caused by smoking cigarettes. They should also believe that inhaling cigarette smoke is a protection. To believe either is, however, to run the risk of failing to recognize and therefore failing to prevent other and more genuine causes...

6. The 1959 Booklet [CP 276A]: The Inhaling Question

In 1959, Fisher's earlier writings on the subject were reprinted in a booklet entitled "Smoking: The Cancer Controversy. Some Attempts to Assess the Evidence", together with a preface and an article on the inhaling issue [CP 276A]. The preface bemoans the mathematical nature of statistical education without regard to the place of statistics in scientific research. "If, indeed, the statistical departments ... were ... clarifying and confirming, in the future research workers who came within their influence, an understanding of the art of examining observational data, the fallacious conclusions drawn, from a simple association, could scarcely have been made the basis of a terrifying propaganda."

In the new piece on the effect of inhaling (recall Tables 2-4), Fisher further analyses the 1948-49 data, now available in finer detail (see Table 9).

Table 9: Cancer Incidence Rates for Inhalers and Non-Inhalers

		Maximum Daily Cigarettes									
		1-4		5-14		15-24		25-49		> 49	
		I	N	I	N	I	N	I	N	I	N
Men	Cancer	7	17	141	67	133	63	96	78	21	24
	Control	17	21	162	80	157	44	74	44	16	7
Women	Cancer	3	3	7	8	7	5	5	3	0	0
	Control	2	10	2	7	6	0	0	0	1	0

(I = Inhaler; N = Non-Inhaler)

Fisher analyzes only the data for men, the number of women being too few in number for reliable analysis. For each of the five two-by-two tables for men "we may ask how many of the inhalers would have shown cancer, if the proportion had been the same as that among non-inhalers". This suggests fixing the wrong margins of the tables, but in computing the expected numbers of inhalers among the cancer patients under the hypothesis of no effect, Fisher correctly fixes the row margins of each 2×2 table. Thus for the typical table,

	I	N
cancer	a	b
control	c	d

,

the expected value corresponding to a is $c(a+b)/(c+d)$. A deficiency is then calculated as $a-c(a+b)/(c+d)$. The results are given in Table 10 with minor corrections to Fisher's values in brackets. In order to assess the significance of the results, Fisher then computes "reduced deficiencies", which are simply the values of $-(ad - bc)/n$. These are convenient since exact standard errors are available for them. Although no explanation is given for the calculation of variance for

Table 10: Comparison of Observed Incidence of Inhalers in the Cancer Group with the Expected Incidence Under No Interaction, Men Only

Cigarettes Per Day	Expected Count $(\frac{a+b}{c+d} \times c)$	Observed Count (a)	Deficiency = Expected - Observed
1-4	10.737	7	3.737
5-14	138.380 (139.24)	142 (141)	-3.620 (-1.76)
15-24	153.095	133	20.095
25-49	109.119	96	13.119
> 49	33.260 (31.30)	21	12.260 (10.30)

the reduced deficiency, the formula used by Fisher is clearly

$$\mathrm{Var}\left(\frac{ad - bc}{n}\right) = \frac{(a+b)\times(a+c)\times(b+d)\times(c+d)}{n^3}$$

in agreement with the formula for chi-square in a 2×2 table. (The exact variance conditional on the marginal totals has $n^2(n-1)$ in place of n^3.) Table 11 gives his computed values of reduced deficiency and associated sampling variances, with minor numerical corrections in parentheses.

Table 11: Reduced Deficiencies and Associated Sampling Variances Derived from Tables 9 and 10

Cigarettes Per Day	Reduced Deficiency $(\frac{bc-ad}{n})$	Sampling Variance
1-4	2.290	3.49
5-14	-1.947 (-0.947)	14.60
15-24	10.174	19.54
25-49	5.301	17.10
> 49	3.485	3.75 (3.78)
Total	20.299 (20.303)	68.48 (68.61)
Standard error of total reduced deficiency = 8.274 (8.277)		

In conclusion, Fisher writes:

> ... No particular importance need be attached to the test of significance. It disproves at about the 1 percent level the hypothesis that inhalers and non-inhalers have the same cancer incidence. Even equality would be a fair knock-out for the theory that smoke in the lung causes cancer...

Should not these workers have let the world know, not only that they have discovered the cause of lung cancer (cigarettes), but also that they had discovered the means of its prevention (inhaling cigarette smoke)? How had the M.R.C. the heart to withhold this information from the thousands who would otherwise die of lung cancer?

Those who refuse the jump from association to causation in the case of cigarette smoking will not be tempted to take it in the case of inhaling; but the M.R.C. and its Statistical Research Unit think the argument is valid in the first case. Can they refuse to admit it in the second?

7. Postscript

Finally, we look briefly at the "Report of the Advisory Committee to the Surgeon General" in Smoking and Health (1964). This report did not comment on the effect of inhaling as did Fisher; neither did it refer to the Doll and Hill study. However, the report did recognize a certain plausibility in the constitutional hypothesis (genetic theory) but regarded the arguments behind the hypothesis as labored and the evidence as either tenuous or actually lacking. It was felt that even if it were true, the constitutional hypothesis would play a minor role relative to that of cigarette smoking.

References

Cornfield, J., et al. (1959). "Smoking and Lung Cancer: Recent Evidence and a Discussion of Some Questions," Journal of the National Cancer Institute, 22, 173-203.

"Dangers of Cigarette-Smoking" (1957). Editorial of British Medical Journal, 1, June 29, 1518-1520.

Doll, R. and B. Hill (1950). "Smoking and Carcinoma of the Lung," British Medical Journal, 2, September 30, 739-748.

Doll, R. and B. Hill (1952). "A Study of the Aetiology of Carcinoma of the Lung," British Medical Journal, 2, December 13, 1272-1286.

Doll, R. and B. Hill (1956). "Lung Cancer and Other Causes of Death in Relation to Smoking," British Medical Journal, 2, November 10, 1071-1081.

Friberg, L., L. Kaij, S. Dencker, and E. Jonsson (1969). "Smoking Habits of Monozygotic and Dizygotic Twins," British Medical Journal, 1, April 25, 1090-1092.

"Geminus" (1958). "It Seems to Me," New Scientist, 4, 440.

Smoking and Health (1964). Report of the Advisory Committee to the Surgeon General of the Public Health Service. Public Health Service Publication No. 1103.

"Tobacco Smoking and Cancer of the Lung" (1957). Statement by the Medical Research Council, British Medical Journal, 1, June 29, 1523-1524.

APPENDIX

PUBLICATIONS OF R.A. FISHER

A selected list of books by R.A. Fisher dealing with statistical topics is included in this Appendix. The list of Collected Papers here follows the listing given in the *Collected Papers of R.A. Fisher* (edited by J.H. Bennett, The University of Adelaide, South Australia: Coudrey Offset Press, 1974), and uses the same numbering scheme as was used there.

SELECT LISTING OF BOOKS BY R.A. FISHER

[SMRW] Statistical Methods for Research Workers. Edinburgh: Oliver and Boyd, 1925, 1928, 1930, 1932, 1934, 1936, 1941, 1944, 1946, 1950, 1954, 1958, 1970; New York: Hafner, 1971, 1973. Also published in French, German, Italian, Japanese, Spanish, and Russian.

[DOE] The Design of Experiments. Edinburgh: Oliver and Boyd, 1935, 1937, 1942, 1949, 1951, 1960, 1966. Also published in Italian, Japanese, and Spanish.

[ST] Statistical Tables for Biological, Agricultural and Medical Research (with F. Yates). Edinburgh: Oliver and Boyd, 1938, 1943, 1948, 1953, 1957, 1963. Also published in Spanish and Portuguese.

[SMSI] Statistical Methods and Scientific Inference. Edinburgh: Oliver and Boyd, 1956, 1959; New York: Hafner, 1973.

COLLECTED PAPERS OF R.A. FISHER

1912 1. On an absolute criterion for fitting frequency curves. *Messeng. Math.*, 41, 155-160.

1913 2. Applications of vector analysis to geometry. *Messeng. Math.*, 42, 161-178.

1914 3. Some hopes of a eugenist. *Eugen. Rev.*, 5, 309-315.

1915 4. Frequency distribution of the values of the correlation coefficient in samples from an indefinitely large population. *Biometrika*, 10, 507-521.

5. (With C.S. Stock). Cuénot on preadaptation: a criticism. *Eugen. Rev.*, 7, 46-61.

6. The evolution of sexual preference. *Eugen. Rev.*, 7, 184-192.

1916 7. Biometrika. *Eugen. Rev.*, 8, 62-64.

1917 8. Positive eugenics. *Eugen. Rev.*, 9, 206-212.

1918 9. The correlation between relatives on the supposition of Mendelian inheritance. *Trans. Roy. Soc. Edinb.*, 52, 399-433.

10. The causes of human variability. *Eugen. Rev.*, 10, 213-220.

1919 11. The genesis of twins. *Genetics*, 4, 489-499.

1920 12. A mathematical examination of the methods of determining the accuracy of an observation by the mean error, and by the mean square error. *Mon. Not. Roy. Ast. Soc.*, 80, 758-770.

13. Review of *Inbreeding and Outbreeding* (E.M. East and D.F. Jones). *Eugen. Rev.*, 12, 116-119.

1921 14. On the "probable error" of a coefficient of correlation deduced from a small sample. *Metron*, 1, 3-32.

15. Studies in crop variation. I. An examination of the yield of dressed grain from Broadbalk. *J. Agric. Sci.*, 11, 107-135.

16. Some remarks on the methods formulated in a recent article on the quantitative analysis of plant growth. *Ann. Appl. Biol.*, 7, 367-372.

17. Review of *The Relative Value of the Processes Causing Evolution* (A.L. and A.C. Hagedoorn). *Eugen. Rev.*, 13, 467-470.

1922 18. On the mathematical foundations of theoretical statistics. *Phil. Trans., A*, 222, 309-368.

19. On the interpretation of χ^2 from contingency tables, and the calculation of P. *J. Roy. Stat. Soc.*, 85, 87-94.

20. The goodness of fit of regression formulae, and the distribution of regression coefficients. J. Roy. Stat. Soc., 85, 597-612.

21. (With W.A. Mackenzie). The correlation of weekly rainfall. Q. J. Roy. Met Soc., 48, 234-242.

22. (With H.G. Thornton and W.A. Mackenzie). The accuracy of the plating method of estimating the density of bacterial populations. Ann. Appl. Biol., 9, 325-359.

23. Statistical appendix to a paper by J. Davidson on Biological studies of Aphis rumicis. Ann. Appl. Biol., 9, 142-145.

24. On the dominance ratio. Proc. Roy. Soc. Edinb., 42, 321-341.

25. The systematic location of genes by means of crossover observations. Amer. Nat., 56, 406-411.

26. Darwinian evolution by mutations. Eugen. Rev., 14, 31-34.

27. New data on the genesis of twins. Eugen. Rev., 14, 115-117.

28. The evolution of the conscience in civilised communities. Eugen. Rev., 14, 190-193.

29. Contribution to a discussion on the inheritance of mental qualities, good and bad. Eugen. Rev., 14, 210-213.

1923

30. Note on Dr. Burnside's recent paper on errors of observation. Proc. Camb. Phil. Soc., 21, 655-658.

31. Statistical tests of agreement between observation and hypothesis. Economica, 3, 139-147.

32. (With W.A. Mackenzie). Studies in crop variation. II. The manurial response of different potato varieties. J. Agric. Sci., 13, 311-320.

33. Paradoxical rainfall data. Nature, 111, 465.

1924

34. The conditions under which χ^2 measures the discrepancy between observation and hypothesis. J. Roy. Stat. Soc., 87, 442-450.

35. The distribution of the partial correlation coefficient. Metron, 3, 329-332.

36. On a distribution yielding the error functions of several well known statistics. Proc. In. Cong. Math., Toronto, 2, 805-813.

37. The influence of rainfall on the yield of wheat at Rothamsted. Phil. Trans., B, 213, 89-142.

38. A method of scoring coincidences in tests with playing cards. Proc. Soc. Psych. Res., 34, 181-185.

39. (With S. Odén). The theory of the mechanical analysis of sediments by means of the automatic balance. Proc. Roy. Soc. Edinb., 44, 98-115.

40. The elimination of mental defect. Eugen. Rev., 16, 114-116.

41. The biometrical study of heredity. Eugen. Rev., 16, 189-210.

1925 42. Theory of statistical estimation. Proc. Camb. Phil. Soc., 22, 700-725.

43. Applications of "Student's" distribution. Metron, 5, 90-104.

44. Expansion of "Student's" integral in powers of n^{-1}. Metron, 5, 109-120.

45. (With P.R. Ansell). Note on the numerical evaluation of a Bessel function derivative. Proc. Lond. Math. Soc., Series 2, 24, liv-lvi.

46. Sur la solution de l'équation intégrale de M.V. Romanovsky. C. R. Acad. Sci., Paris, 181, 88-89.

47. The resemblance between twins, a statistical examination of Lauterbach's measurements. Genetics, 10, 569-579.

1926 48. The arrangement of field experiments. J. Min. Agric. G. Br., 33, 503-513.

49. Bayes' theorem and the fourfold table. Eugen. Rev., 18, 32-33.

50. On the random sequence. Q.J. Roy. Met. Soc., 52, 250.

51. On the capillary forces in an ideal soil: correction of formulae given by W.B. Haines. J. Agric. Sci., 16, 492-503.

52. (With E.B. Ford). Variability of species. Nature, 118, 515-516.

53. Eugenics: Can it solve the problem of decay of civilisations? Eugen. Rev., 18, 128-136.

54. Modern eugenics. Sci. Prog., 21, 130-136; Eugen. Rev., 18, 231-236.

55. Periodical health surveys. J. State Med., 34:446-449.

1927 56. (With J. Wishart). On the distribution of the error of an interpolated value, and on the construction of tables. Proc. Camb. Phil. Soc., 23, 912-921.

57. (With T. Eden). Studies in crop variation. IV. The experimental determination of the value of top dressings with cereals. J. Agric. Sci., 17, 548-562.

58. (With H.G. Thornton). On the existence of daily changes in the bacterial numbers in American Soil. Soil Sci., 23, 253-259.

59. On some objections to mimicry theory -- statistical and genetic. Trans. Roy. Ent. Soc. Lond., 75, 269-278.

60. The actuarial treatment of official birth records. Eugen. Rev., 19, 103-108.

1928 61. The general sampling distribution of the multiple correlation coefficient. Proc. Roy. Soc., A, 121, 654-673.

62. On a property connecting the χ^2 measure of discrepancy with the method of maximum likelihood. Atti. Cong. Int. Mat., Bologna, 6, 95-100.

63. (With L.H.C. Tippett). Limiting forms of the frequency distribution of the largest or smallest member of a sample. Proc. Camb. Phil. Soc., 24, 180-190.

64. Further note on the capillary forces in an ideal soil. J. Agric. Sci., 18, 406-410.

65. (With T.N. Hoblyn). Maximum- and minimum-correlation tables in comparative climatology. Geogr. Ann., 10, 267-281.

66. Correlation coefficients in meteorology. Nature, 121, 712.

67. The effect of psychological card preferences. Proc. Soc. Psych. Res., 38, 269-271.

68. The possible modification of the response of the wild type to recurrent mutations. Am. Nat., 62, 115-126.

69. Two further notes on the origin of dominance. Am. Nat., 62, 571-574.

70. Triplet children in Great Britain and Ireland. Proc. Roy. Soc., B, 102, 286-311.

71. (With B. Balmukand). The estimation of linkage from the offspring of selfed heterozygotes. J. Genet., 20, 79-92.

72. (With E.B. Ford). The variability of species in the Lepidoptera, with reference to abundance and sex. Trans. Roy. Entom. Soc. Lond., 76, 367-379.

73. The differential birth rate: new light on causes from American figures. Eugen. Rev., 20, 183-184.

1929

74. Moments and product moments of sampling distributions. Proc. Lond. Math. Soc., Series 2, 30, 199-238.

75. Tests of significance in harmonic analysis. Proc. Roy. Soc., A, 125, 54-59.

76. The sieve of Eratosthenes. Math. Gaz., 14, 564-566.

77. A preliminary note on the effect of sodium silicate in increasing the yield of barley. J. Agric. Sci., 19, 132-139.

78. (With T. Eden). Studies in crop variation. VI. Experiments on the response of the potato to potash and nitrogen. J. Agric. Sci., 19, 201-213.

79. The statistical method in psychical research. Proc. Soc. Psych. Res., 39, 189-192.

80. Statistics and biological research. Nature, 124, 266-267.

81. The evolution of dominance: a reply to Professor Sewall Wright. Am. Nat., 63:553-556.

82. The over-production of food. Realist, 1, 45-60.

1930 83. The moments of the distribution for normal samples of measures of departure from normality. Proc. Roy. Soc., A, 130, 16-28.

84. Inverse probability. Proc. Camb. Phil. Soc., 26, 528-535.

85. (With J. Wishart). The arrangement of field experiments and the statistical reduction of the results. Imp. Bur. Soil Sci. Tech. Comm., 10, 23 pp.

86. The distribution of gene ratios for rare mutations. Proc. Roy. Soc. Edinb., 50, 205-220.

87. The evolution of dominance in certain polymorphic species. Am. Nat., 64, 385-406.

88. Mortality amongst plants and its bearing on natural selection. Nature, 125, 972-973.

89. Note on a tri-colour (mosaic) mouse. J. Genet., 23, 77-81.

1931 90. (With J. Wishart). The derivation of the pattern formulae of two-way partitions from those of simpler patterns. Proc. Lond. Math. Soc., Series 2, 33, 195-208.

91. The sampling error of estimated deviates, together with other illustrations of the properties and applications of the integrals and derivatives of the normal error function. Brit. Assn. Math. Tab., 1, xxvi-xxxv (3rd ed., xxviii-xxxvii, 1951).

92. (With S. Bartlett). Pasteurised and raw milk. Nature, 127, 591-592.

93. The evolution of dominance. Biol. Rev., 6, 345-368.

94. The biological effects of family allowances. Family Endowment Chronicle, 1, 21-25.

1932 95. Inverse probability and the use of likelihood. Proc. Camb. Phil. Soc., 28, 257-261.

96. (With F.R. Immer and O. Tedin). The genetical interpretation of statistics of the third degree in the study of quantitative inheritance. Genetics, 17, 107-124.

97. The evolutionary modification of genetic phenomena. Proc. 6th Int. Congr. Genet., 1, 165-172.

98. The bearing of genetics on theories of evolution. Sci. Prog., 27, 273-287.

99. The social selection of human fertility. The Herbert Spencer Lecture, 32 pp. Oxford: Clarendon Press.

100. Family allowances in the contemporary economic situation. Eugen. Rev., 24, 87-95.

101. Inheritance of acquired characters. Nature, 130, 579.

1933 102. The concepts of inverse probability and fiducial probability referring to unknown parameters. Proc. Roy. Soc., A, 139, 343-348.

103. The contributions of Rothamsted to the development of the science of statistics. Annual Report Rothamsted Experimental Station, 43-50.

104. On the evidence against the chemical induction of melanism in Lepidoptera. Proc. Roy. Soc., B, 112, 407-416.

105. Selection in the production of ever-sporting stocks. Ann. Bot., 47, 727-733.

106. Number of Mendelian factors in quantitative inheritance. Nature, 131, 400-401.

107. Contribution to a discussion on mortality among young plants and animals. Proc. Linn. Soc. Lond., 145, 100-101.

1934 108. Two new properties of mathematical likelihood. Proc. Roy. Soc., A, 144, 285-307.

109. Probability, likelihood and quantity of information in the logic of uncertain inference. Proc. Roy. Soc., A, 146, 1-8.

110. (With F. Yates). The 6 × 6 Latin squares. Proc. Camb. Phil. Soc., 30, 492-507.

111. Randomisation, and an old enigma of card play. Math. Gaz., 18, 294-297.

112. Appendix to a paper by H.G. Thornton and P.H.H. Gray on the numbers of bacterial cells in field soils. Proc. Roy. Soc., B, 115, 540-542.

113. The effect of methods on ascertainment upon the estimation of frequencies. Ann. Eugen., 6, 13-25.

114. The amount of information supplied by records of families as a function of the linkage in the population sampled. Ann. Eugen., 6, 66-70.

115. The use of simultaneous estimation in the evaluation of linkage. Ann. Eugen., 6, 71-76.

116. Some results of an experiment on dominance in poultry, with special reference to polydactyly. Proc. Linn. Soc. Lond., 147, 71-81.

117. Crest and hernia in fowls due to a single gene without dominance. Science, 80, 288-289.

118. (With C. Diver). Crossing-over in the land snail Cepaea nemoralis. Nature, 133, 834-835.

119. Professor Wright on the theory of dominance. Am. Nat., 68, 370-374.

120. The children of mental defectives. In the Report of Departmental Cttee. on Sterilisation, 60-74, H.M.S.O.

121. Indeterminism and natural selection. Philos. Sci., 1, 99-117.

122. Adaptation and mutations. School Sci. Rev., 15, 294-301.

1935 123. The mathematical distributions used in the common tests of significance. Econometrica, 3, 353-365.

124. The logic of inductive inference. J. Roy. Stat. Soc., 98, 39-54.

125. The fiducial argument in statistical inference. Ann. Eugen., 6, 391-398.

126. The case of zero survivors in probit assays. Ann. Appl. Biol., 22, 164-165.

127. Statistical tests. Nature, 136, 474.

128. Contribution to a discussion of J. Neyman's paper on statistical problems in agricultural experimentation. J. Roy. Stat. Soc., Suppl., 2, 154-157, 173.

129. Contribution to a discussion of F. Yates' paper on complex experiments. J. Roy. Stat. Soc., Suppl., 2, 229-231.

130. On the selective consequences of East's theory of heterostylism in Lythrum. J. Genet., 30, 369-382.

131. The detection of linkage with "dominant" abnormalities. Ann. Eugen., 6, 187-201.

132. The detection of linkage with recessive abnormalities. Ann. Eugen., 6, 339-351.

133. The sheltering of lethals. Am. Nat., 69, 446-455.

134. The inheritance of fertility: Dr. Wagner-Manslau's tables. Ann. Eugen., 6, 225-251.

135. Dominance in poultry. Philos. Trans., B, 225, 197-226.

136. Eugenics, academic and practical. Eugen. Rev., 27, 95-100.

1936 137. Uncertain inference. Proc. Am. Acad. Arts Sci., 71, 245-258.

138. The use of multiple measurements in taxonomic problems. Ann. Eugen., 7, 179-188.

139. (With S. Barbacki). A test of the supposed precision of systematic arrangements. Ann. Eugen., 7, 189-193.

140. The half-drill strip system agricultural experiments. Nature, 138, 1101.

141. "The coefficient of racial likeness" and the future of craniometry. J. Roy. Anthropol. Inst., 66, 57-63.

142. Heterogeneity of linkage data for Friedreich's ataxia and the spontaneous antigens. Ann. Eugen., 7, 17-21.

143. Tests of significance applied to Haldane's data on partial sex linkage. Ann. Eugen., 7, 87-104.

144. Has Mendel's work been rediscovered? Ann. Sci., 1, 115-137.

145. (With K. Mather). A linkage test with mice. Ann. Eugen., 7, 265-280.

146. (With K. Mather). Verification in mice of the possibility of more than fifty per cent recombination. Nature, 137, 362-363.

147. The measurement of selective intensity. Proc. Roy. Soc., B, 121, 58-62.

1937 148. (With E.A. Cornish). Moments and cumulants in the specification of distributions. Rev. Inst. Int. Stat., 5, 307-322.

149. Professor Karl Pearson and the method of moments. Ann. Eugen., 7, 308-318.

150. (With B. Day). The comparison of variability in populations having unequal means. An example of the analysis of covariance with multiple dependent and independent variates. Ann. Eugen., 7, 333-348.

151. On a point raised by M.S. Bartlett on fiducial probability. Ann. Eugen., 7, 370-375.

152. The wave of advance of advantageous genes. Ann. Eugen., 7, 355-369.

153. The relation between variability and abundance shown by the measurements of the eggs of British nesting birds. Proc. Roy. Soc., B, 122, 1-26.

154. (With H. Gray). Inheritance in man: Boas's data studied by the method of analysis of variance. Ann. Eugen., 8, 74-93.

1938 155. The statistical utilization of multiple measurements. Ann. Eugen., 8, 376-386.

156. Quelques remarques sur l'estimation en statistique. Biotypologie, 6, 153-158.

157. On the statistical treatment of the relation between sea-level characteristics and high-altitude acclimatization. Proc. Roy. Soc., B, 126, 25-29.

158. The mathematics of experimentation. Nature, 142, 442-443.

159. Presidential address, Indian statistical conference. Sankhyā, 4, 14-17.

160. Comment on D. McGregor's note on the distribution of the three forms of Lythrum salicaria. Ann. Eugen., 8, 177.

161. Dominance in poultry: feathered feet, rose comb, internal pigment and pile. Proc. Roy. Soc., B, 125, 25-48.

1939 162. The comparison of samples with possibly unequal variances. Ann. Eugen., 9, 174-180.

163. The sampling distribution of some statistics obtained from non-linear equations. Ann. Eugen., 9, 238-249.

164. A note on fiducial inference. Ann. Math. Stat., 10, 383-388.

165. "Student". Ann. Eugen., 9, 1-9.

166. The precision of the product formula for the estimation of linkage. Ann. Eugen., 9, 50-54.

167. Selective forces in wild populations of *Paratettix texanus*. *Ann. Eugen.*, 9, 109-122.

168. Stage of development as a factor influencing the variance in the number of offspring, frequency of mutants and related quantities. *Ann. Eugen.*, 9, 406-408.

169. (With G.L. Taylor). Blood groups in Great Britain. *Brit. Med. J.*, 2, 826.

170. (With E.B. Ford and J. Huxley). Taste-testing the anthropoid apes. *Nature*, 144, 750.

171. (With J. Vaughan). Surnames and blood-groups. *Nature*, 144, 1047.

172. The Galton Laboratory. *Times*, 29 September; *Science*, 90, 436.

1940 173. On the similarity of the distributions found for the test of significance in harmonic analysis, and in Steven's problem in geometrical probability. *Ann. Eugen.*, 10, 14-17.

174. An examination of the different possible solutions of a problem in incomplete blocks. *Ann. Eugen.*, 10, 52-75.

175. The precision of discriminant functions. *Ann. Eugen.*, 10, 422-429.

176. The estimation of the proportion of recessives from tests carried out on a sample not wholly unrelated. *Ann. Eugen.*, 10, 160-170.

177. (With W.H. Dowdeswell and E.B. Ford). The quantitative study of populations in the Lepidoptera. I. *Polyommatus icarus*. *Ann. Eugen.*, 10, 123-136.

178. (With K. Mather). Non-lethality of the mid factor in *Lythrum salicaria*. *Nature*, 146, 521.

179. (With G.L. Taylor). Scandinavian influence in Scottish ethnology. *Nature*, 145, 590.

180. The Galton Laboratory. *Science*, 91, 44-45.

1941 181. The asymptotic approach to Behrens's integral, with further tables for the *d* test of significance. *Ann. Eugen.*, 11, 141-172.

182. The negative binomial distribution. *Ann. Eugen.*, 11, 182-187.

183. The interpretation of experimental four-fold tables. *Science*, 94, 210-211.

184. The theoretical consequences of polyploid inheritance for the mid style form of *Lythrum salicaria*. *Ann. Eugen.*, 11, 31-38.

185. Average excess and average effect of a gene substitution. *Ann. Eugen.*, 11, 53-63.

1942 186. New cyclic solutions to problems in incomplete blocks. *Ann. Eugen.*, 11, 290-299.

187. Completely orthogonal 9 × 9 squares - a correction. *Ann. Eugen.*, 11, 402-403.

188. The likelihood solution of a problem in compounded probabilities. *Ann. Eugen.*, 11, 306-307.

189. The theory of confounding in factorial experiments in relation to the theory of groups. *Ann. Eugen.*, 11, 341-353.

190. Some combinatorial theorems and enumerations connected with the numbers of diagonal types of a Latin square. *Ann. Eugen.*, 11, 395-401.

191. The polygene concept. *Nature*, 150, 154.

192. (With K. Mather). Polyploid inheritance in *Lythrum salicaria*. *Nature*, 150, 430.

1943 193. A theoretical distribution for the apparent abundance of different species. *J. Anim. Ecol.*, 12, 54-58.

194. Note on Dr. Berkson's criticism of tests of significance. *J. Am. Stat. Assn.*, 38, 103-104.

195. (With W.R.G. Atkins). The therapeutic use of vitamin C. *J. Roy. Army Med. Corps*, 83, 251-252.

196. (With K. Mather). The inheritance of style length in *Lythrum salicaria*. *Ann. Eugen.*, 12, 1-23.

197. (With J.A. Fraser Roberts). A sex difference in blood-group frequencies. *Nature*, 151, 640-641.

198. The birthrate and family allowances. *Agenda*, 2, 124-133.

1944 199. (With S.B. Holt). The experimental modification of dominance in Danforth's short-tailed mutant mice. *Ann. Eugen.*, 12, 102-120.

200. Allowance for double reduction in the calculation of genotype frequencies with polysomic inheritance. *Ann. Eugen.*, 12, 169-171.

201. (With R.R. Race and G.L. Taylor). Mutation and the Rhesus reaction. *Nature*, 153, 106.

1945 202. A system of confounding for factors with more than two alternatives, giving completely orthogonal cubes and higher powers. *Ann. Eugen.*, 12, 283-290.

203. The logical inversion of the notion of the random variable. *Sankhya*, 7, 129-132.

204. Recent progress in experimental design. In *L'application du calcul des probabilités*, 19-31. *Proc. Int. Inst. Intell. Coop., Geneva* (1939).

205. A new test for 2 × 2 tables. *Nature*, 156, 388.

206. (With L. Martin). The hereditary and familial aspects of exophthalmic goitre and nodular goitre. *Q.J. Med.*, 14, 207-219.

1946 207. Testing the difference between two means of observations of unequal precision. *Nature*, 158, 713.

208. A system of scoring linkage data, with special reference to the pied factors in mice. *Am. Nat.*, 80, 568-578.

209. (With R.R. Race). Rh gene frequencies in Britain. *Nature*, 157, 48-49.

210. The fitting of gene frequencies to date on *Rhesus* reactions. *Ann. Eugen.*, 13, 150-155.

1947 211. The analysis of covariance method for the relation between a part and the whole. *Biometrics*, 3, 65-68.

212. Development of the theory of experimental design. *Proc. Int. Statist. Conf.*, 3, 434-439.

213. The theory of linkage in polysomic inheritance. *Philos. Trans., B*, 233, 55-87.

214. The *Rhesus* factor: a study in scientific method. *Am. Sci.*, 35, 95-102, 113.

215. Note on the calculation of the frequencies of *Rhesus* allelomorphs. *Ann. Eugen.*, 13, 223-224.

216. The science of heredity. *Listener*, 37, 662-663.

217. The renaissance of Darwinism. *Listener*, 37, 1001, 1009; *Parents' Review*, 58, 183-187.

218. (With V.C. Martin). Spontaneous occurrence in *Lythrum salicaria* of plants duplex for the short-style gene. *Nature*, 160, 541.

219. (With E.B. Ford). The spread of a gene in natural conditions in a colony of the moth *Panaxia dominula*. *Heredity*, 1, 143-174.

220. Number of self-sterility alleles. *Nature*, 160, 797-798.

221. (With M.F. Lyon and A.R.G. Owen). The sex chromosome in the house mouse. *Heredity*, 1, 355-365.

1948 222. Conclusions fiduciaires. *Ann. Inst. Henri Poincare*, 10, 191-213.

223. (With D. Dugué). Un résultat assez inattendu d'arithmétique des lois de probabilité. *C.R. Acad. Sci., Paris*, 227, 1205-1206.

224. Biometry. *Biometrics*, 4, 217-219.

225. A quantitative theory of genetic recombination and chiasma formation. *Biometrics*, 4, 1-9.

226. (With G.D. Snell). A twelfth linkage group of the house mouse. *Heredity*, 2, 271-273.

227. (With V.C. Martin). Genetics of style-length in *Oxalis*. *Nature*, 162, 533.

228. Modern genetics. *Brit. Sci. News*, 1, 2-4.

229. What sort of man is Lysenko? *Soc. Freed. Sci., Occasl. Pamp.*, 9, 6-9; *Listener*, 40, 874-875.

1949 230. A biological assay of tuberculins. Biometrics, 5, 300-316.

231. Note on the test of significance for differential viability in frequency data from a complete three-point test. Heredity, 3, 215-219.

232. (With W.H. Dowdeswell and E.B. Ford). The quantitative study of populations in the Lepidoptera. 2. Maniola jurtina. Heredity, 3, 67-84.

233. A preliminary linkage test with agouti and undulated mice. Heredity, 3, 229-241.

234. A theoretical system of selection for homostyle Primula. Sankhyā, 9, 325-342.

235. The linkage problem in a tetrasomic wild plant, Lythrum salicaria. Proc. 8th Int. Congr. Genet. (suppl. to Hereditas), 225-233.

1950 236. The significance of deviations from expectation in a Poisson series. Biometrics, 6, 17-24.

237. A class of enumerations of importance in genetics. Proc. Roy. Soc., B, 136, 509-520.

238. Gene frequencies in a cline determined by selection and diffusion. Biometrics, 6, 353-361.

239. (With E.B. Ford). The "Sewall Wright effect". Heredity, 4, 117-119.

240. Polydactyly in mice. Nature, 165, 407, 796.

241. Creative Aspects of Natural Law. The Eddington Memorial Lecture. 23 pp. Cambridge: Cambridge University Press.

1951 242. Statistics, in Scientific Thought in the Twentieth Century. A.E. Heath, Ed. London: Watts, pp. 31-55.

243. Standard calculations for evaluating a blood-group system. Heredity, 5, 95-102.

244. A combinatorial formulation of multiple linkage tests. Nature, 167, 520.

245. Limits to intensive production in animals. Brit. Agric. Bull., 4, 217-218.

246. (With L. Martin). The hereditary and familial aspects of toxic nodular goitre (secondary thyrotoxicosis). Q.J. Med., 20, 293-297.

1952 247. Sequential experimentation. Biometrics, 8, 183-187.

248. Statistical methods in genetics. The Bateson Lecture, 1951. Heredity, 6, 1-12.

1953 249. Dispersion on a sphere. Proc. Roy. Soc., A, 217, 295-305.

250. Note on the efficient fitting of the negative binomial. Biometrics, 9, 197-199.

251. The expansion of statistics. J. Roy. Stat. Soc., A, 116, 1-6; Am. Sci., 42, 275-282, 293.

252. Population genetics. The Croonian Lecture, 1953. Proc. Roy. Soc., B, 141, 510-523.

253. The variation in strength of the human blood group P. Heredity, 7, 81-89.

254. The linkage of polydactyly with leaden in the house mouse. Heredity, 7, 91-95.

255. (With W. Landauer). Sex differences of crossing-over in close linkage. Am. Nat., 87, 116.

1954 256. The analysis of variance with various binomial transformations. Biometrics, 10, 130-139.

257. Contribution to a discussion of a paper on interval estimation by M.A. Creasy. J. Roy. Statist. Soc., B, 16, 212-213.

258. Retrospect of the criticisms of the theory of natural selection. In Evolution as a Process, J.S. Huxley, A.C. Hardy, and E.B. Ford, Eds. London: Allen and Unwin, pp. 84-98.

259. A fuller theory of "junctions" in inbreeding. Heredity, 8, 187-197.

260. The experimental study of multiple crossing over. Proc. 9th Int. Congr. Genet., Caryologia, 6, Suppl., 227-231.

1955 261. Statistical methods and scientific induction. J. Roy. Stat. Soc., B, 17, 69-78.

262. (With V.C. Fyfe). Double reduction at the rosy, or pink, locus in Lythrum salicaria. Nature, 176, 1176.

263. Science and Christianity. Friend, 113 (42), 2 pp.

1956 264. On a test of significance in Pearson's Biometrika tables (no. 11). J. Roy. Stat. Soc., B, 18, 56-60.

265. (With M.J.R. Healy). New tables of Behrens' test of significance. J. Roy. Stat. Soc., B, 18, 56-60.

266. Blood groups and population genetics. Proc. 1st Int. Congr. Human Genet., Acta Genet., 6, 507-509.

1957 267. The underworld of probability. Sankhyā, 18, 201-210.

268. Comment on the notes by Newman, Bartlett and Welch in this Journal. (18, 288-302). J. Roy. Stat. Soc., B, 19, 179.

269. Dangers of cigarette-smoking. Brit. Med. J., 2, 43.

270. Dangers of cigarette-smoking. Brit. Med. J., 2, 297-298.

271. Methods in human genetics. Proc. 1st Int. Congr. Human Genet., Acta Genet., 7, 7-10.

1958 272. The nature of probability. Centennial Rev., 2, 261-274.

273. Mathematical probability in the natural sciences. *Proc. 18th Int. Congr. Pharmaceut. Sci.*; *Metrika*, 2, 1-10; *Technometrics*, 1, 21-29; *La Scuola in Azione*, 20, 5-19.

274. Cigarettes, cancer, and statistics. *Centennial Rev.*, 2, 151-166.

275. Lung cancer and cigarettes? *Nature*, 182, 108.

276. Cancer and smoking. *Nature*, 182, 596.

276a. *Smoking: The Cancer Controversy. Some Attempts to Assess the Evidence.* Edinburgh: Oliver and Boyd.

277. Polymorphism and natural selection. *Bull. Inst. Int. Stat.*, 36, 284-289; *J. Ecol.*, 46, 289-293.

278. The discontinuous inheritance. *Listener*, 60, 85-87.

1959 279. Natural selection from the genetical standpoint. *Aust. J. Sci.*, 22, 16-17.

280. An algebraically exact examination of junction formation and transmission in parent-offpsring inbreeding. *Heredity*, 13, 179-186.

1960 281. (With E.A. Cornish). The percentile points of distributions having known cumulants. *Technometrics*, 2, 209-225.

282. Scientific thought and the refinement of human reasoning. *J. Oper. Res. Soc. Japan*, 3, 1-10.

283. On some extensions of Bayesian inference proposed by Mr. Lindley. *J. Roy. Stat. Soc., B*, 22, 299-301.

1961 284. Sampling the reference set. *Sankhyā*, 23, 3-8.

285. The weighted mean of two normal samples with unknown variance ratio. *Sankhyā*, 23, 103-114.

286. Possible differentiation in the wild population of *Oenothera organensis*. *Aust. J. Biol. Sci.*, 14, 76-78.

287. A model for the generation of self-sterility alleles. *J. Theoret. Biol.*, 1, 411-414.

1962 288. The simultaneous distribution of correlation coefficients. *Sankhyā*, 24, 1-8.

289. Some examples of Bayes' method of the experimental determination of probabilities a priori. *J. Roy. Stat. Soc., B*, 24, 118-124.

290. The place of the design of experiments in the logic of scientific inference. *Colloq. Int. Cent. Nat. Recherche Scientifique, Paris*, 110, 13-19; *La Scuola in Azione*, 9, 33-42 (in Italian).

291. Confidence limits for a cross-product ratio. *Aust. J. Stat.*, 4, 41.

292. Enumeration and classification in polysomic inheritance. *J. Theoret. Biol.*, 2, 309-311.

293. Self-sterility alleles: a reply to Professor D. Lewis. J. Theoret. Biol., 2, 309-311.

294. The detection of a sex difference in recombination values using double heterozygotes. J. Theoret. Biol., 3, 509-513.